菠萝种质资源图谱（上册）
Pineapple Germplasm Resources Map (Volume 1)

◎ 陆新华　张秀梅　刘胜辉　吴青松　主编

 中国农业科学技术出版社

图书在版编目（CIP）数据

菠萝种质资源图谱 . 上册 / 陆新华等主编 . -- 北京：中国农业科学技术出版社，2023.10
ISBN 978-7-5116-6470-9

Ⅰ.①菠… Ⅱ.①陆… Ⅲ.①菠萝－种质资源－中国－图集 Ⅳ.① S668.3-64

中国国家版本馆 CIP 数据核字（2023）第 200835 号

责任编辑　史咏竹　白　净
责任校对　马广洋
责任印制　姜义伟　王思文

出 版 者	中国农业科学技术出版社
	北京市中关村南大街 12 号　　邮编：100081
电　　话	（010）82105169（编辑室）　（010）82109702（发行部）
	（010）82109709（读者服务部）
网　　址	https://castp.caas.cn
经 销 者	各地新华书店
印 刷 者	北京地大彩印有限公司
开　　本	170 mm×240 mm　1/16
印　　张	13.5
字　　数	229 千字
版　　次	2023 年 10 月第 1 版　2023 年 10 月第 1 次印刷
定　　价	98.00 元

◆◆◆ 版权所有 · 侵权必究 ◆◆◆

《菠萝种质资源图谱（上册）》
编委会

作者名单

主　　编　　陆新华　张秀梅　刘胜辉　吴青松
副 主 编　　黄炳钰　林文秋　高玉尧　李川玲
参编人员　　（以姓氏笔画为序）
　　　　　　　井敏敏　付　琼　朱祝英　孙光明
　　　　　　　孙伟生　杜丽清　杨玉梅　汪佳维
　　　　　　　姚艳丽　贺军军　夏　瑞

作者单位

中国热带农业科学院南亚热带作物研究所
农业农村部湛江菠萝种质资源圃
国家热带植物种质资源库
热带作物生物育种全国重点实验室
农业农村部热带果树生物学重点实验室
海南省菠萝种质创新与利用工程技术研究中心
国家重要热带作物工程技术研究中心—菠萝研发部
华南农业大学园艺学院

Pineapple Germplasm Resources Map (Volume 1)
Editorial Board

The List of Authors

Editors in Chief Lu Xinhua, Zhang Xiumei, Liu Shenghui, Wu Qingsong

Associate Editors in Chief Huang Bingyu, Lin Wenqiu, Gao Yuyao, Li Chuanling

Editors (Sort by the strokes of the surnames in Chinese)

　　　　Jing Minmin, Fu Qiong, Zhu Zhuying, Sun Guangming, Sun Weisheng,
　　　　Du Liqing, Yang Yumei, Wang Jiawei, Yao Yanli, He Junjun, Xia Rui

Affiliations of the Authors

South Subtropical Crop Research Institute, Chinese Academy of Tropical Agricultural Sciences

Germplasm Repository of Pineapple (*Ananas comosus*) Zhanjiang City, Ministry of Agriculture and Rural Affairs

National Tropical Plants Germplasm Resource Center

National Key Laboratory for Tropical Crop Breeding

Key Laboratory of Tropical Fruit Biology, Ministry of Agriculture and Rural Affairs

Hainan Provincial Engineering Research Center for Pineapple Germplasm Innovation and Utilization

Pineapple R&D Department, National Center of Important Tropical Crops Engineering and Technology Research

College of Horticulture, South China Agricultural University

本书资助项目：农业农村部农垦局项目"菠萝、香蕉、澳洲坚果、荔枝等南亚热带作物种质资源保护与利用"（项目编号：A120202）；国家重点研发计划课题"菠萝种质资源精准评价与基因发掘"（项目编号：2019YFD1000505）；广东省农业农村厅项目"岭南特色水果新品种区域示范及推广"（项目编号：440000220000000035430）。

This book was supported by projects of Agricultural Reclamation Bureau, Ministry of Agriculture and Rural Affairs (A120202), National Key R & D Program of China (2019YFD1000505), Agricultural and Rural Department of Guangdong Province (440000220000000035430).

本书采集图片资料的田间工作操作地点：农业农村部湛江菠萝种质资源圃。

The field operation location for collecting image materials in this book is the Germplasm Repository of Pineapple (*Ananas comosus*) Zhanjiang City, Ministry of Agriculture and Rural Affairs.

本书采集图片资料及统计性状过程中各品种菠萝定植时间：2021年2月。

During the process of collecting image materials and statistical traits in this book, various varieties of pineapple were planted in February 2021.

目 录

| 第一章 | 菠萝种质资源遗传多样性图谱 | 1 |

| 第二章 | 国内菠萝种质资源 | 19 |

　　无刺卡因 20
　　开英1号 24
　　台农4号 28
　　台农17号 32
　　台农19号 36
　　台农21号 40
　　杂233 44
　　黑菠萝 48
　　神湾菠萝 52
　　上海2号 56
　　红皮菠萝 60

| 第三章 | 国外菠萝种质资源 | **65** |

　　MD-2 66
　　金冬蜜 70
　　鲜优菠萝 74
　　马若奇 78
　　麦格雷戈 82
　　碧武里1号 86
　　碧武里2号 90
　　普吉 94
　　新普吉 98
　　Tradsrithong 102

Contents

Chapter 1　Genetic Diversity Map of Pineapple Germplasm Resources ·········· 1

Chapter 2　Pineapple Germplasm Resources in China ·········· 19

　　Smooth Cayenne ·········· 21
　　Cayenne No. 1 ·········· 25
　　Tainung No. 4 ·········· 29
　　Tainung No. 17 ·········· 33
　　Tainung No. 19 ·········· 37
　　Tainung No. 21 ·········· 41
　　Hybrid 233 ·········· 45
　　Black Pineapple ·········· 49
　　Shenwan Pineapple ·········· 53
　　Shanghai No. 2 ·········· 57
　　Red skin Pineapple ·········· 61

Chapter 3　Pineapple Germplasm Resources from Abroad ·········· 65

　　MD-2 ·········· 67
　　Golden Winter Sweet ·········· 71
　　Fresh Premium Pineapple ·········· 75
　　Maroochy ·········· 79
　　MacGregor ·········· 83
　　Phetchaburi No. 1 ·········· 87
　　Phetchaburi No. 2 ·········· 91
　　Phuket ·········· 95
　　New Phuket ·········· 99
　　Tradsrithong ·········· 103

无刺卡因（巴冲） 106
泰国菠萝 110
Baro Rothchild 114
C180 118
Charlotte Rothschild 122
Kallara 土种 126
詹姆斯皇后 130
印度皇后 134
印度引未知名 1 138
印度引未知名 2 142
N36 146
沙捞越 150
Josapine 154
观赏类绿果 158
马来西亚引未知名 1 162
DL1 166
DL3 170
DN1 174
DN2 178
DN5 182
JPZ 186
越南引皇后 1 号 190
越南引无刺卡因 2 号 194
肯尼亚引无刺卡因 1 号 198
法国野生种 202

参考文献 206

Smooth Cayenne (Pakchong) ... 107
Thailand Pineapple ... 111
Baro Rothchild ... 115
C180 ... 119
Charlotte Rothschild ... 123
Kallara Local ... 127
James Queen ... 131
Indian Queen ... 135
Unknown cultivar No. 1 from Indian ... 139
Unknown cultivar No. 2 from Indian ... 143
N36 ... 147
Sarawak ... 151
Josapine ... 155
Ornamental Green Fruit ... 159
Unknown cultivar No. 1 from Malaysia ... 163
DL1 ... 167
DL3 ... 171
DN1 ... 175
DN2 ... 179
DN5 ... 183
JPZ ... 187
Queen No. 1 from Vietnam ... 191
Smooth Cayenne No. 2 from Vietnam ... 195
Smooth Cayenne No. 1 from Kenya ... 199
Wild Type from France ... 203

Reference ... **206**

第一章
菠萝种质资源遗传多样性图谱

Chapter 1
Genetic Diversity Map of Pineapple Germplasm Resources

菠萝［*Ananas comosus* (L.) Merr.］是凤梨科（Bromeliaceae）凤梨属（*Ananas* Merr.）多年生草本植物。菠萝原产于南美洲，1493年被欧洲人首次发现，后来推广至世界各地，现广泛分布于热带和亚热带地区。菠萝果实风味独特、香气怡人、营养丰富，可鲜食，亦可制成罐头、果汁或用于提取菠萝蛋白酶等，是世界四大热带水果（香蕉、菠萝、荔枝和杧果）之一，具有重要的经济价值。我国最早引进菠萝是在16世纪末，迄今已有400多年历史。菠萝现已发展成为我国热区重要的经济果品之一，主要分布在广东、海南、广西、云南、台湾等省区。中国热带农业科学院南亚热带作物研究所从2003年起引入菠萝，目前收集保存150多份种质资源，种植于农业农村部湛江菠萝种质资源圃内。本图谱仅对圃内保存的有代表性的菠萝种质资源的植物学性状、农艺性状和品质性状等进行了描述。

不同品种菠萝种质资源在植株姿态、冠芽特征、冠芽外形、冠芽叶刺、叶片彩带、叶片叶刺、叶刺生长方向等方面具有一定的差异。植株姿态有直立（夹角≥80°）、开张（40°≤夹角＜80°）、匍匐（夹角＜40°）；冠芽特征有单冠芽、双冠芽、多冠芽、单小冠芽（冠芽高度小于果体高度的1/2）；冠芽外形有椭圆形、圆柱形、圆锥形、扇形、喇叭形、其他形态；冠芽叶刺状态分为光滑无刺、部分叶缘有刺、叶尖有刺、全缘有刺；叶片着生姿态分为直立、开张、平展、下垂；叶片颜色有淡绿/绿色、绿色带黄色斑纹、绿色带紫红色斑纹、暗红色、浅紫/紫红色、深紫/暗紫红色、其他颜色等；叶片彩带状态分为无、两侧、中央；叶片叶刺分布状态有光滑无刺、仅少量分布在叶尖处、仅少量且无规律地分布在叶缘处、较多且无规律地分布在叶缘处、布满整个叶缘；叶刺生长方向有向上顺生、向上顺生与向下倒生兼备；叶刺密度有少量/稀疏（密度≤1枚/cm）、中度（1枚/cm＜密度＜3枚/cm）、很多/密集（密度≥3枚/cm）等类型。

第一章 菠萝种质资源遗传多样性图谱
Chapter 1 Genetic Diversity Map of Pineapple Germplasm Resources

Pineapple [*Ananas comosus* (L.) Merr.] belonging to the genus *Ananas* of the family Bromeliaceae, is a perennial herbaceous plant. Pineapple was originated from South America. It was first discovered by Europeans in 1493. Later, pineapple was spread to many other parts of the world and now it is widely distributed in the tropical and subtropical areas. Pineapple fruit has rich nutrition with unique flavors and a pleasant aroma, and is usually consumed as fresh fruit or canned products, it can also be used to produce juices or bromelains. Pineapple is one of the four most important tropical fruits (banana, pineapple, litchi and mango) in the world and it has an essential economic value. The earliest introduction of pineapple to China was at the end of the 16th century and now it has more than 400 years of history here. Currently, pineapple has become one of principal economic fruit in tropical regions of China, being mainly distributed in Guangdong, Hainan, Guangxi, Yunan, Taiwan, and other places. Since 2003, South Subtropical Crops Research Institute (SSCRI) of Chinese Academy of Tropical Agricultural Sciences (CATAS) has introduced its first pineapple germplasm, and now more than 150 pineapple accessions are collected and conserved in Germplasm Repository of Pineapple (*Ananas comosus*) Zhanjiang City, Ministry of Agriculture and Rural Affairs. This book introduces some typical pineapple germplasm resources conserved in the germplasm repository and their botanical, agronomic and quality characteristics, etc.

Pineapple germplasm resources possess rich diversity in the aspects of their plant postures, crown bud characters, crown bud shape, crown bud spines, leaf ribbons, leaf spines, leaf spine growth directions, etc. The plant posture of pineapple can be either upright (angle$\geqslant$80°), spreading (40°$\leqslant$angle$<$80°) or procumbent (angle$<$40°). They could have either a single crown bud, a double crown bud, a multiple crown bud or a single small crown bud (the height of the crown bud is less than half of the height of the fruit body). The crown bud can be either ellipsoid, cylindrical, conical, fan-shaped, trumpet shaped, etc. The crown bud leaf has smooth margin without spines, partly margin with spines, tip with spines or overall margins with spines. The leaf posture can be either erect, spreading, expanding or droopy. Leaf color can be green/light green, green with yellow stripe, green with purplish red stripe, dark red, light purple/purplish red, dark purple/dark purplish red, etc. Some varieties do not have leaf ribbon but others have it in the middle or on both sides of the leaf. The characteristics of leaf spines vary from one variety to another: smooth margins without spines, few spines distributing at the leaf tips, few spines irregularly distributing at the leaf margins, many spines regularly distributing at the leaf margins, spines regularly distributing at both sides of the leaf margins, or

在菠萝花序上的小花开放盛期，花瓣开张状态有微开、半张开、张开；按照花瓣间是否重叠及其状态，花冠形态有旋转状、覆瓦状；苞片边缘状态有锯齿状、光滑。

菠萝果实具有丰富的遗传多样性，按果实形状可分为圆台形/方形、（近）圆球形、圆筒形、长圆筒形、圆锥形、长圆锥形、其他；按未成熟果实果皮颜色可分为银绿色、淡绿/绿色、暗绿色、暗墨绿色、淡黄/黄绿色、浅红/粉红/橙红/浅褐色、红色、红色中略显紫色、暗紫红色、紫色、蓝紫色等；按成熟果实果皮颜色可分为绿色、黄色、带绿斑、暗黄/深黄色、亮黄/淡黄色、金黄/鲜黄色等；按果基形状可分为平、弧形、突起；按果顶形状可分为平、浑圆、钝圆、尖圆；果颈可分为无颈、有颈；按果眼外观可分为扁平或微凹、微隆起、突起/隆起；按果眼深度可分为深（深度≥1.2 cm）、较深（1.1 cm≤深度＜1.2 cm）、较浅（0.9 cm≤深度＜1.1 cm）和浅（深度＜0.9 cm）；按果眼排列方式可分为左旋、右旋、其他类型；按果实底部着生的果瘤可分为无、少（果瘤数1～2个）、多（果瘤数≥3个）；按果实小果可分为不可剥离、可剥离；按果肉颜色可分为白色、奶油色、淡黄色、黄色、金黄色、深黄色、橙黄色等；按果实香味可分为无香、清香/微香、芳香、极香/浓香等；按果肉风味可分为浓甜、清甜、甜酸、酸、酸甜、极酸、微香甜、微涩等；按果肉质地可分为滑、脆/爽脆、粗糙。按果实成熟特性分为特早熟（成熟期≤60 d）、早熟（60 d＜成熟期＜80 d）、中熟（80 d≤成熟期＜100 d）、晚熟（≥100 d）等；根据菠萝果实第一次采收至最后一次采收之间所持续的天数，成熟期的一致性可分为一致（相差天数≤5 d）、基本一致（5 d＜相差天数＜15 d）、不一致（相差天数≥15 d）；按产量特性可分为丰产（产量≥40 t/hm^2）、一般产量（30 t/hm^2≤产量＜40 t/hm^2）、低产（产量＜30 t/hm^2）。

spines covering the entire leaf margins. The leaf spines are either antrorse or both antrorse and retrorse at the same time. Spine density can be sparse (density≤1/cm), medium (1/cm<density<2/cm), or many/condensed (density≥3/cm), etc.

During the full-blooming stage, petals of florets in an inflorescence can be slightly open, half open, or fully open. According to whether the petals overlap and their state, the corolla morphology can be classified into either contorted or imbricate. The margins of bract can be either serrated or smooth.

The pineapple fruit presents a rich genetic diversity. The fruit shape can be either truncated cone/square, (near) spherical, cylindrical, long cylindrical, conical, long conical, pyriform, etc. The immature fruit peel can be silver green, light green/green, dark green, dark blackish green, light yellow/yellowish green, light red/candy pink/orange red/light brown, red, red with slight purple, violet red, purple, blue purple, etc. In addition, the mature fruit peel can be green, yellow with green stripe, dark yellow/deep yellow, bright yellow/light yellow, golden yellow/vivid yellow, etc. Fruit base can be flat, curved or protrusive. Fruit top can be flat, perfectly round, bluntly round or sharply round. Some pineapples show obvious fruit neck, others have null. Fruit eyes (fruitlets) can be flat, slightly concave, slightly embossed, or convex/embossed. The fruit eyes depth can be deep (depth≥1.2 cm), relatively deep (1.1 cm≤depth<1.2 cm), relatively shallow (0.9 cm≤depth<1.1 cm) or shallow (depth≤0.9 cm). The fruit eyes arrangement can be levorotation, dextrorotation or other type. Some varieties do not have fruit tumors on the bottom of the fruit, but others have few (1~2) or many (≥3) fruit tumors. Some fruitlets can be easily separated but others can not. Flesh color can be white, cream, pale yellow, yellow, golden yellow, deep yellow, orange, etc. Some fruits give off no scent but many other fruits send out faint/slight scent, sweet fragrance, or even strong/extreme aroma. Fruit flavour can be strong sweet, mildly sweet, sweet-sour, sour, sour-sweet, extreme sour, slightly fragrant and sweet, astringent, etc. The flesh texture can be smooth, crisp/crunchy or coarse. Fruit ripening characteristics can be divided into extremely early mature (mature period≤60 d), early mature (60 d<mature period<80 d), medium mature (80 d≤mature period<100 d) or late mature (mature period≥100 d). According to the duration days between the first and the last harvest of fruit, the maturity consistency can be uniform (duration of days≤5 d), basically uniform (5 d<duration of days<15 d) or nonuniform (duration of days≥15 d). The fruit yield characteristics can be productive (production≥40 t/hm^2), normal level (30 t/hm^2≤production<40 t/hm^2) or low (production<30 t/hm^2).

植株姿态　Plant posture

直立
Upright

开张
Spreading

匍匐
Procumbent

第一章　菠萝种质资源遗传多样性图谱
Chapter 1　Genetic Diversity Map of Pineapple Germplasm Resources

冠芽特征　Crown bud characteristic

单冠芽
Single

双冠芽
Double

多冠芽
Multiple

单小冠芽
Single small

冠芽外形 Crown bud shape

椭圆形
Ellipsoid

圆柱形
Cylindrical

圆锥形
Conical

扇形
Fan-shaped

喇叭形
Trumpet shaped

第一章 菠萝种质资源遗传多样性图谱
Chapter 1　Genetic Diversity Map of Pineapple Germplasm Resources

冠芽叶刺　Leaf spines of crown bud

光滑无刺
Absent

部分叶缘有刺
Part leaf margin with spines

叶尖有刺
Leaf tip with spines

全缘有刺
Overall leaf margin with spines

叶片着生姿态　Leaf posture

竖直
Erect

开张
Spreading

平展
Expanding

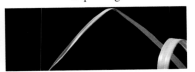

下垂
Droopy

叶片彩带状态 Leaf ribbon

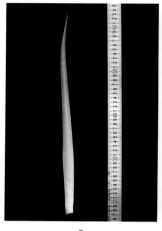
无
Absent

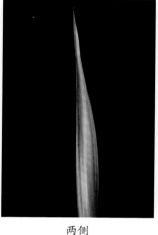
两侧
Both sides

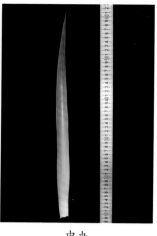
中央
Middle

叶片叶刺分布状态 Distribution of leaf spines

无
Absent

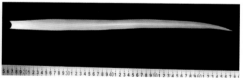
仅少量分布在叶尖处
Few spines distributing at the leaf tip

仅少量且无规律地分布在叶缘处
Few spines irregularly distributing at the leaf margin

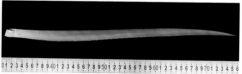
较多且无规律地分布在叶缘处
Many spines regularly distributing at the leaf margin

布满整个叶缘
Spines covering the entire leaf margin

第一章 菠萝种质资源遗传多样性图谱
Chapter 1 Genetic Diversity Map of Pineapple Germplasm Resources

叶刺生长方向　Growth direction of leaf spines

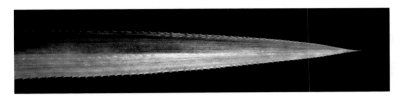

向上顺生
Antrorse

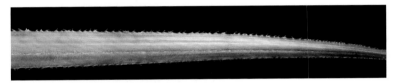

向上顺生与向下倒生兼备
Both antrorse and retrorse

叶刺密度　Density of leaf spines

稀疏
Sparse

中度
Medium

密集
Condensed

花瓣开张状态 State of petal opening

微开
Slightly open

半张开
Half open

张开
Fully open

花冠形态 Corolla morphology

旋转状
Contorted

覆瓦状
Imbricate

第一章　菠萝种质资源遗传多样性图谱
Chapter 1　Genetic Diversity Map of Pineapple Germplasm Resources

果颈　Fruit neck

无
Absent

有
Present

果眼外观　Fruit eyes exterior

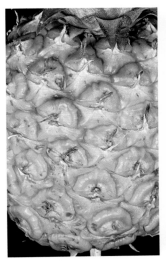

扁平或微凹
Flat or slightly concave

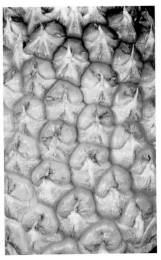

微隆起
Slightly embossed

突起/隆起
Convex/ embossed

果眼排列方式　Fruit eyes arrangement

左旋
Levorotation

右旋
Dextrorotation

其他类型
Other type

果实底部着生的果瘤　Fruit tumors on the bottom of the fruit

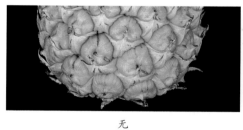
无
Absent

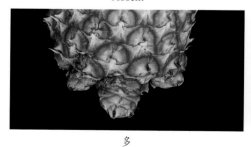
多
Many

少
Few

第一章 菠萝种质资源遗传多样性图谱
Chapter 1 Genetic Diversity Map of Pineapple Germplasm Resources

果肉颜色　Flesh color

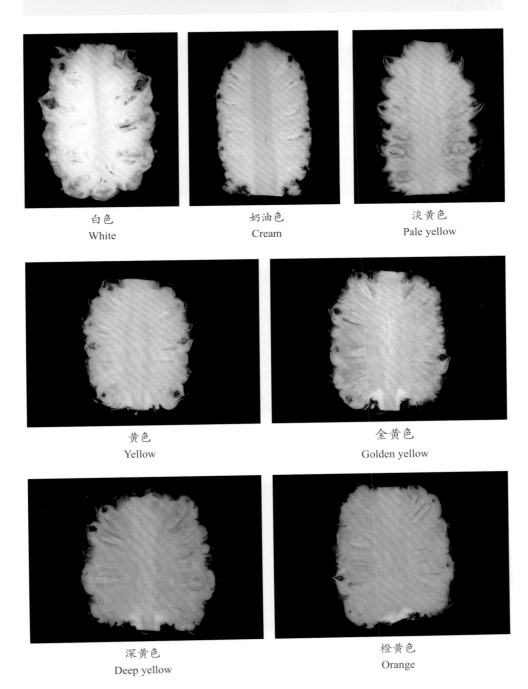

白色　White

奶油色　Cream

淡黄色　Pale yellow

黄色　Yellow

金黄色　Golden yellow

深黄色　Deep yellow

橙黄色　Orange

第二章
国内菠萝种质资源

Chapter 2
Pineapple Germplasm Resources in China

无刺卡因

编号：ACCESSION000018
种质名称：无刺卡因
原产地：不详
资源类型：国内收集
主要用途：鲜食或加工
种质来源地：2007年中国热带农业科学院南亚热带作物研究所从广东省收集
植株姿态：开张
定植期：2021年2月中旬（湛江，余同）
现红期：2022年3月中旬（自然果，余同）
营养生长期：约390 d
初花期：2022年3月下旬
花开放时间：约16 d
成熟期：2022年7月中下旬
果实发育期：约115 d
果实形状：长圆柱
单果质量：1021 g
纵径：13.0 cm
横径：9.8 cm
果形指数：1.3
未成熟果实果皮颜色：银绿色
成熟果实果皮颜色：暗黄/深黄色
果颈：无
果基：平
果顶：钝圆
果实小果能否剥离：不可剥离
果眼外观：扁平或微凹
果眼深度：深
果眼排列方式：左旋或右旋
果瘤：多
果肉颜色：淡黄色
果实香味：清香/微香
果实风味：酸甜
果肉质地：滑
果实外观综合评价：中
果实品质综合评价：好
果肉可溶性固形物含量：16.5%
果实成熟特性：晚熟
成熟期的一致性：不一致
丰产性：一般

Smooth Cayenne

Code: ACCESSION000018
Name: Smooth Cayenne
Place of origin: unknown
Resource type: domestically collected variety
Main ways of consumption: fresh fruits or processed products
Material resource: South Subtropical Crop Research Institute of Chinese Academy of Tropical Agricultural Science (SSCRI, CATAS, the same below) collected from Guangdong Province in 2007
Plant posture: spreading
Planting time: mid-February in 2021 (Zhanjiang, the same below)
Open heart stage: mid-March in 2022 (natural fruiting, the same below)
Vegetative growth period: approx. 390 d
Initial flowering time: late March in 2022
Flowering period: approx. 16 d
Fruit maturation time: mid-to late-July in 2022
Fruit developing period: approx. 115 d
Fruit shape: long cylindrical
Single fruit weight: 1021g
Longitudinal diameter: 13.0 cm
Transverse diameter: 9.8 cm
Fruit shape index: 1.3
Immature fruit peel color: silver green
Mature fruit peel color: dark yellow/deep yellow
Fruit neck: absent
Fruit base: flat
Fruit top: bluntly round
Fruitlets adhesion situation: inseparable
Fruit eyes appearance: flat or slightly concave
Fruit eyes depth: deep
Fruit eyes arrangement: levorotation or dextrorotation
Fruit tumors: many
Flesh color: pale yellow
Fruit aroma: faint/slight scent
Fruit flavour: sour-sweet
Flesh texture: smooth
Fruit appearance comprehensive evaluation: medium
Fruit quality comprehensive evaluation: excellent
Soluble solids content in flesh: 16.5%
Fruit ripening characteristics: late mature
Maturity consistency: nonuniform
Productivity: normal level

菠萝种质资源图谱 （上册）
Pineapple Germplasm Resources Map (Volume 1)

植株
Plant

现红
Open heart

花
Flower

花序
Inflorescence

第二章　国内菠萝种质资源
Chapter 2　Pineapple Germplasm Resources in China

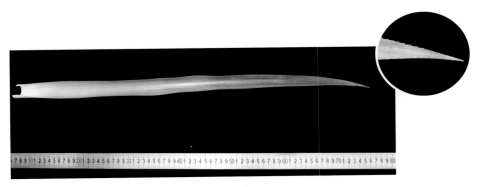

叶片
Leaf

带冠芽果
Fruit with crown bud

冠芽叶刺
Leaf spines in crown bud

果实
Fruit

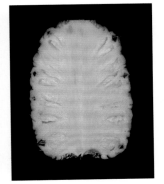

果实纵切
Fruit longitudinal section

果实横切
Fruit transverse section

开英 1 号

编号：ACCESSION000078
种质名称：开英 1 号，又名台农 1 号
原产地：不详
资源类型：国内收集
主要用途：鲜食或加工
种质来源地：2012 年中国热带农业科学院南亚热带作物研究所从海南省收集
植株姿态：开张
定植期：2021 年 2 月上旬
现红期：2022 年 3 月上旬
营养生长期：约 390 d
初花期：2022 年 3 月下旬
花开放时间：约 18 d
成熟期：2022 年 7 月中旬
果实发育期：约 110 d
果实形状：圆筒形
单果质量：1283 g
纵径：16.2 cm
横径：11.2 cm
果形指数：1.4

未成熟果实果皮颜色：银绿色
成熟果实果皮颜色：暗黄/深黄色
果颈：无
果基：弧形
果顶：平
果实小果能否剥离：可剥离
果眼外观：扁平或微凹
果眼深度：较浅
果眼排列方式：左旋或右旋
果瘤：多
果肉颜色：淡黄色
果实香味：清香/微香
果实风味：酸甜
果肉质地：滑
果实外观综合评价：优
果实品质综合评价：好
果肉可溶性固形物含量：17.7%
果实成熟特性：晚熟
成熟期的一致性：不一致
丰产性：丰产

Cayenne No. 1

Code: ACCESSION000078
Name: Cayenne No. 1, also known as Tainung No. 1
Place of origin: unknown
Resource type: domestically collected variety
Main ways of consumption: fresh fruits or processed products
Material resource: SSCRI, CATAS collected from Hainan Province in 2012
Plant posture: spreading
Planting time: early February in 2021
Open heart stage: early March in 2022
Vegetative growth period: approx. 390 d
Initial flowering time: late March in 2022
Flowering period: approx. 18 d
Fruit maturation time: mid-July in 2022
Fruit developing period: approx. 110 d
Fruit shape: cylindrical
Single fruit weight: 1283 g
Longitudinal diameter: 16.2 cm
Transverse diameter: 11.2 cm
Fruit shape index: 1.4
Immature fruit peel color: silver green
Mature fruit peel color: dark /deep yellow
Fruit neck: absent
Fruit base: curved
Fruit top: flat
Fruitlets adhesion situation: separable
Fruit eyes appearance: flat or slightly concave
Fruit eyes depth: relatively shallow
Fruit eyes arrangement: levorotation or dextrorotation
Fruit tumors: many
Flesh color: pale yellow
Fruit aroma: faint/slight scent
Fruit flavour: sour-sweet
Flesh texture: smooth
Fruit appearance comprehensive evaluation: excellent
Fruit quality comprehensive evaluation: good
Soluble solids content in flesh: 17.7 %
Fruit ripening characteristics: late mature
Maturity consistency: nonuniform
Productivity: productive

菠萝种质资源图谱 （上册）
Pineapple Germplasm Resources Map　(Volume 1)

植株
Plant

现红
Open heart

花
Flower

花序
Inflorescence

第二章 国内菠萝种质资源
Chapter 2 Pineapple Germplasm Resources in China

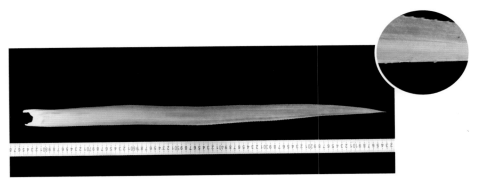

叶片
Leaf

带冠芽果
Fruit with crown bud

冠芽叶刺
Leaf spines in crown bud

果实
Fruit

果实纵切
Fruit longitudinal section

果实横切
Fruit transverse section

台农 4 号

编号：ACCESSION000021

种质名称：台农 4 号，又名剥粒菠萝、手撕菠萝

资源类型：选育品种

主要用途：鲜食或加工

系谱：Smooth Cayenne (♀) × 新加坡种 (♂) 的杂交后代

选育单位：中国台湾农业试验所嘉义农业试验站

植株姿态：开张

定植期：2021 年 2 月中旬

现红期：2022 年 2 月下旬

营养生长期：约 370 d

初花期：2022 年 3 月下旬

花开放时间：约 20 d

成熟期：2022 年 7 月上中旬

果实发育期：约 105 d

果实形状：圆筒形

单果质量：1049 g

纵径：14.2 cm

横径：11.6 cm

果形指数：1.2

未成熟果实果皮颜色：暗墨绿色

成熟果实果皮颜色：深黄至橙色

果颈：有

果基：平

果顶：平

果实小果能否剥离：可剥离

果眼外观：突起 / 隆起

果眼深度：较深

果眼排列方式：右旋

果瘤：无或少

果肉颜色：金黄色

果实香味：清香 / 微香

果实风味：清甜

果肉质地：粗糙

果实外观综合评价：优

果实品质综合评价：好

果肉可溶性固形物含量：17.0%

果实成熟特性：晚熟

成熟期的一致性：基本一致

丰产性：一般

Tainung No. 4

Code: ACCESSION000021
Name: Tainung No. 4, also known as Peeling Pineapple, Hand-peeled Pineapple
Resource type: bred variety
Main ways of consumption: fresh or processing
Lineage: a hybrid between Smooth Cayenne (♀) and Singapore (♂)
Breeding organization: Jiayi Agricultural Experiment Station, Taiwan Agricultural Research Institute, China
Plant posture: spreading
Planting time: mid-February in 2021
Open heart stage: late February in 2022
Vegetative growth period: approx. 370 d
Initial flowering time: late March in 2022
Flowering period: approx. 20 d
Fruit maturation time: early-to-mid July in 2022
Fruit developing period: approx. 105 d
Fruit shape: cylindrical
Single fruit weight: 1049 g
Longitudinal diameter: 14.2 cm
Transverse diameter: 11.6 cm
Fruit shape index: 1.2
Immature fruit peel color: dark blackish green
Mature fruit peel color: deep yellow to orange
Fruit neck: present
Fruit base: flat
Fruit top: flat
Fruitlets adhesion situation: separable
Fruit eyes appearance: convex/ embossed
Fruit eyes depth: relatively deep
Fruit eyes arrangement: dextrorotation
Fruit tumors: absent or few
Flesh color: golden yellow
Fruit aroma: faint/slight scent
Fruit flavour: mildly sweet
Flesh texture: coarse
Fruit appearance comprehensive evaluation: excellent
Fruit quality comprehensive evaluation: good
Soluble solids content in flesh: 17.0%
Fruit ripening characteristics: late mature
Maturity consistency: basically uniform
Productivity: normal level

菠萝种质资源图谱 （上册）
Pineapple Germplasm Resources Map (Volume 1)

植株
Plant

现红
Open heart

花
Flower

花序
Inflorescence

第二章 国内菠萝种质资源
Chapter 2　Pineapple Germplasm Resources in China

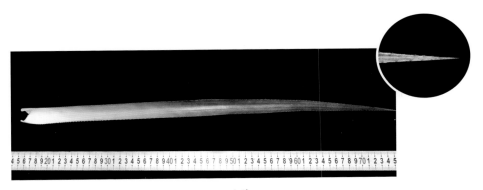

叶片
Leaf

带冠芽果
Fruit with crown bud

冠芽叶刺
Leaf spines in crown bud

果实
Fruit

果实纵切
Fruit longitudinal section

果实横切
Fruit transverse section

台农 17 号

编号：ACCESSION000006
种质名称：台农 17 号，又名金钻菠萝
资源类型：选育品种
主要用途：鲜食或加工
系　谱：Smooth Cayenne（♀）× Queen（♂）的杂交后代
选育单位：中国台湾农业试验所嘉义农业试验站
植株姿态：开张
定植期：2021 年 2 月上旬
现红期：2022 年 3 月上旬
营养生长期：约 390 d
初花期：2022 年 3 月下旬
花开放时间：约 21 d
成熟期：2022 年 6 月中下旬
果实发育期：约 85 d
果实形状：圆筒形
单果质量：1188 g
纵径：15.3 cm
横径：11.0 cm

果形指数：1.4
未成熟果实果皮颜色：暗墨绿色
成熟果实果皮颜色：金黄色
果颈：有
果基：弧形
果顶：平
果实小果能否剥离：可剥离
果眼外观：突起 / 隆起
果眼深度：较深
果眼排列方式：左旋或其他
果瘤：少或多
果肉颜色：金黄 / 鲜黄色
果实香味：清香 / 微香
果实风味：甜酸
果肉质地：滑
果实外观综合评价：中
果实品质综合评价：好
果肉可溶性固形物含量：18.7%
果实成熟特性：中熟
成熟期的一致性：不一致
丰产性：丰产

Tainung No. 17

Code: ACCESSION000006

Name: Tainung No. 17, also known as Jinzuan Pineapple

Resource type: bred variety

Main ways of consumption: fresh or processing

Lineage: a hybrid between Smooth Cayenne (♀) and Queen (♂)

Breeding organization: Jiayi Agricultural Experiment Station, Taiwan Agricultural Research Institute, China

Plant posture: spreading

Planting time: early February in 2021

Open heart stage: early March in 2022

Vegetative growth period: approx. 390 d

Initial flowering time: late March in 2022

Flowering period: approx. 21 d

Fruit maturation time: mid-to-late June in 2022

Fruit developing period: approx. 85 d

Fruit shape: cylindrical

Single fruit weight: 1188 g

Longitudinal diameter: 15.3 cm

Transverse diameter: 11.0 cm

Fruit shape index: 1.4

Immature fruit peel color: dark blackish green

Mature fruit peel color: golden yellow

Fruit neck: present

Fruit base: curved

Fruit top: flat

Fruitlets adhesion situation: separable

Fruit eyes appearance: convex/ embossed

Fruit eyes depth: relatively deep

Fruit eyes arrangement: dextrorotation or other

Fruit tumors: few or many

Flesh color: golden yellow/vivid yellow

Fruit aroma: faint/slight scent

Fruit flavour: sweet-sour

Flesh texture: smooth

Fruit appearance comprehensive evaluation: excellent

Fruit quality comprehensive evaluation: medium

Soluble solids content in flesh: 18.7 %

Fruit ripening characteristics: medium mature

Maturity consistency: nonuniform

Productivity: productive

菠萝种质资源图谱 （上册）
Pineapple Germplasm Resources Map　(Volume 1)

植株
Plant

现红
Open heart

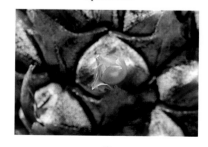

花
Flower

花序
Inflorescence

第二章 国内菠萝种质资源
Chapter 2 Pineapple Germplasm Resources in China

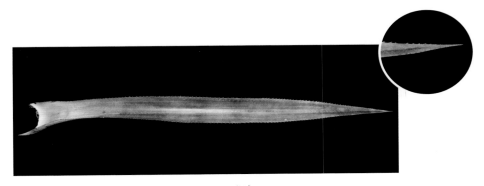

叶片
Leaf

带冠芽果
Fruit with crown bud

冠芽叶刺
Leaf spines in crown bud

果实
Fruit

果实纵切
Fruit longitudinal section

果实横切
Fruit transverse section

台农 19 号

编号：ACCESSION000001
种质名称：台农 19 号，又名蜜宝菠萝
资源类型：选育品种
主要用途：鲜食或加工
系谱：Smooth Cayenne（♀）× Rough（♂）的杂交后代
选育单位：中国台湾农业试验所嘉义农业试验站
植株姿态：开张
定植期：2021 年 2 月下旬
现红期：2022 年 3 月中旬
营养生长期：约 390 d
初花期：2022 年 4 月上旬
花开放时间：约 20 d
成熟期：2022 年 6 月下旬至 8 月上旬
果实发育期：约 100 d
果实形状：圆筒形
单果质量：664 g
纵径：12.3 cm
横径：10.1 cm

果形指数：1.2
未成熟果实果皮颜色：淡绿色
成熟果实果皮颜色：暗黄 / 深黄色
果颈：无
果基：突起
果顶：钝圆
果实小果能否剥离：不可剥离
果眼外观：微隆起
果眼深度：较浅
果眼排列方式：左旋或右旋
果瘤：无
果肉颜色：淡黄色
果实香味：清香 / 微香
果实风味：甜酸
果肉质地：滑
果实外观综合评价：中
果实品质综合评价：中
果肉可溶性固形物含量：12.1%
果实成熟特性：中晚熟
成熟期的一致性：不一致
丰产性：低产

Tainung No. 19

Code: ACCESSION000001
Name: Tainung No. 19, also known as Mibao Pineapple
Resource type: bred variety
Main ways of consumption: fresh or processing
Lineage: a hybrid between Smooth Cayenne (♀) and Rough (♂)
Breeding organization: Jiayi Agricultural Experiment Station, Taiwan Agricultural Research Institute, China
Plant posture: spreading
Planting time: late February in 2021
Open heart stage: mid-March in 2022
Vegetative growth period: approx. 390 d
Initial flowering time: early April in 2022
Flowering period: approx. 20 d
Fruit maturation time: from late June to early August in 2022
Fruit developing period: approx. 100 d
Fruit shape: cylindrical
Single fruit weight: 664 g
Longitudinal diameter: 12.3 cm
Transverse diameter: 10.1 cm
Fruit shape index: 1.2

Immature fruit peel color: light green
Mature fruit peel color: dark yellow/deep yellow
Fruit neck: absent
Fruit base: protrusive
Fruit top: bluntly round
Fruitlets adhesion situation: inseparable
Fruit eyes appearance: slightly embossed
Fruit eyes depth: relatively shallow
Fruit eyes arrangement: levorotation or dextrorotation
Fruit tumors: absent
Flesh color: pale yellow
Fruit aroma: faint/slight scent
Fruit flavour: sweet-sour
Flesh texture: smooth
Fruit appearance comprehensive evaluation: medium
Fruit quality comprehensive evaluation: medium
Soluble solids content in flesh: 12.1 %
Fruit ripening characteristics: medium-late mature
Maturity consistency: nonuniform
Productivity: low

菠萝种质资源图谱 （上册）
Pineapple Germplasm Resources Map (Volume 1)

植株
Plant

现红
Open heart

花
Flower

花序
Inflorescence

第二章 国内菠萝种质资源
Chapter 2　Pineapple Germplasm Resources in China

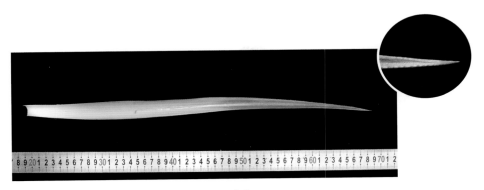

叶片
Leaf

带冠芽果
Fruit with crown bud

冠芽叶刺
Leaf spines in crown bud

果实
Fruit

果实纵切
Fruit longitudinal section

果实横切
Fruit transverse section

台农 21 号

编号：ACCESSION000039
种质名称：台农 21 号，又名黄金菠萝
资源类型：选育品种
主要用途：鲜食或加工
系谱：C64-4-117（♀）× C64-2-56（♂）的杂交后代
选育单位：中国台湾农业试验所嘉义农业试验站
植株姿态：开张
定植期：2021 年 2 月下旬
现红期：2022 年 2 月上旬
营养生长期：约 345 d
初花期：2022 年 3 月下旬
花开放时间：约 24 d
成熟期：2022 年 6 月下旬
果实发育期：约 90 d
果实形状：圆筒形
单果质量：562 g
纵径：10.6 cm
横径：9.7 cm
果形指数：1.1
未成熟果实果皮颜色：暗绿色
成熟果实果皮颜色：金黄色
果颈：无
果基：平
果顶：浑圆
果实小果能否剥离：可剥离
果眼外观：微隆起
果眼深度：深
果眼排列方式：左旋或右旋
果瘤：无
果肉颜色：金黄色
果实香味：清香
果实风味：清甜
果肉质地：脆 / 爽脆
果实外观综合评价：优
果实品质综合评价：优
果肉可溶性固形物含量：20.4%
果实成熟特性：中熟
成熟期的一致性：一致
丰产性：低产

Tainung No. 21

Code: ACCESSION000039
Name: Tainung No. 21, also known as Golden Pineapple
Resource type: bred variety
Main ways of consumption: fresh or processing
Lineage: a hybrid between C64-4-117 (♀) and C64-2-56 (♂)
Breeding organization: Jiayi Agricultural Experiment Station, Taiwan Agricultural Research Institute, China
Plant posture: spreading
Planting time: late February in 2021
Open heart stage: early February in 2022
Vegetative growth period: approx. 345 d
Initial flowering time: late March in 2022
Flowering period: approx. 24 d
Fruit maturation time: late June in 2022
Fruit developing period: approx. 90 d
Fruit shape: cylindrical
Single fruit weight: 562 g
Longitudinal diameter: 10.6 cm
Transverse diameter: 9.7 cm
Fruit shape index: 1.1
Immature fruit peel color: dark green
Mature fruit peel color: golden yellow
Fruit neck: absent
Fruit base: flat
Fruit top: perfectly round
Fruitlets adhesion situation: separable
Fruit eyes appearance: slightly embossed
Fruit eyes depth: deep
Fruit eyes arrangement: levorotation or dextrorotation
Fruit tumors: absent
Flesh color: golden yellow
Fruit aroma: faint scent
Fruit flavour: mildly sweet
Flesh texture: crisp/crunchy
Fruit appearance comprehensive evaluation: excellent
Fruit quality comprehensive evaluation: excellent
Soluble solids content in flesh: 20.4 %
Fruit ripening characteristics: medium mature
Maturity consistency: uniform
Productivity: low

菠萝种质资源图谱 （上册）
Pineapple Germplasm Resources Map (Volume 1)

植株
Plant

现红
Open heart

花
Flower

花序
Inflorescence

第二章　国内菠萝种质资源
Chapter 2　Pineapple Germplasm Resources in China

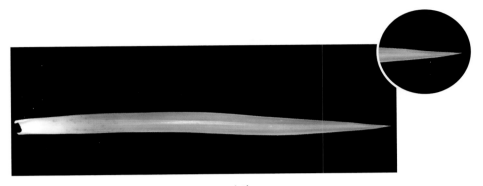

叶片
Leaf

带冠芽果
Fruit with crown bud

冠芽叶刺
Leaf spines in crown bud

果实
Fruit

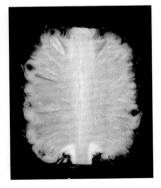

果实纵切
Fruit longitudinal section

果实横切
Fruit transverse section

杂 233

编号：ACCESSION000123
种质名称：杂 233
资源类型：选育品系
主要用途：鲜食或加工
系谱：无刺卡因（♀）× 巴厘（♂）的杂交后代
选育单位：中国热带农业科学院南亚热带作物研究所杂交选育
植株姿态：开张
定植期：2021 年 2 月下旬
现红期：2022 年 2 月下旬
营养生长期：约 365 d
初花期：2022 年 4 月上旬
花开放时间：约 21 d
成熟期：2022 年 6 月中旬至 7 月上旬
果实发育期：约 80 d
果实形状：圆筒形
单果质量：676 g
纵径：12.8 cm
横径：9.7 cm
果形指数：1.3

未成熟果实果皮颜色：银绿色
成熟果实果皮颜色：金黄/鲜黄色
果颈：无
果基：弧形
果顶：平
果实小果能否剥离：可剥离
果眼外观：微隆起
果眼深度：深
果眼排列方式：左旋
果瘤：无或少
果肉颜色：淡黄/黄色
果实香味：清香/微香
果实风味：酸甜
果肉质地：脆/爽脆
果实外观综合评价：优
果实品质综合评价：中
果肉可溶性固形物含量：16.2%
果实成熟特性：早中熟
成熟期的一致性：不一致
丰产性：低产

Hybrid 233

Code: ACCESSION000123
Name: Hybrid 233
Resource type: bred variety
Main ways of consumption: fresh or processing
Lineage: a hybrid between Smooth Cayenne (♀) and Comte de Paris (♂)
Breeding organization: SSCRI, CATAS
Plant posture: spreading
Planting time: late February in 2021
Open heart stage: late February in 2022
Vegetative growth period: approx. 365 d
Initial flowering time: early April in 2022
Flowering period: approx. 21 d
Fruit maturation time: from mid-June to early July in 2022
Fruit developing period: approx. 80 d
Fruit shape: cylindrical
Single fruit weight: 676 g
Longitudinal diameter: 12.8 cm
Transverse diameter: 9.7 cm
Fruit shape index: 1.3
Immature fruit peel color: silver green
Mature fruit peel color: golden yellow/vivid yellow
Fruit neck: absent
Fruit base: curved
Fruit top: flat
Fruitlets adhesion situation: separable
Fruit eyes appearance: slightly embossed
Fruit eyes depth: deep
Fruit eyes arrangement: levorotation
Fruit tumors: absent or few
Flesh color: pale yellow/yellow
Fruit aroma: faint/slight scent
Fruit flavour: sour-sweet
Flesh texture: crisp/crunchy
Fruit appearance comprehensive evaluation: excellent
Fruit quality comprehensive evaluation: medium
Soluble solids content in flesh: 16.2 %
Fruit ripening characteristics: early-medium mature
Maturity consistency: nonuniform
Productivity: low

菠萝种质资源图谱（上册）
Pineapple Germplasm Resources Map (Volume 1)

植株
Plant

现红
Open heart

花
Flower

花序
Inflorescence

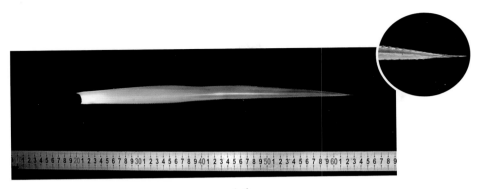

叶片
Leaf

带冠芽果
Fruit with crown bud

冠芽叶刺
Leaf spines in crown bud

果实
Fruit

果实纵切
Fruit longitudinal section

果实横切
Fruit transverse section

黑菠萝

编号：ACCESSION000038
种质名称：黑菠萝
原产地：不详
资源类型：国内收集
主要用途：鲜食或加工
种质来源地：2008年中国热带农业科学院南亚热带作物研究所从云南省瑞丽市收集
植株姿态：开张
定植期：2021年2月中旬
现红期：2022年3月中旬
营养生长期：约390 d
初花期：2022年3月下旬
花开放时间：约22 d
成熟期：2022年6月中旬至7月下旬
果实发育期：约100 d
果实形状：椭圆形
单果质量：629 g
纵径：11.8 cm
横径：9.5 cm
果形指数：1.2

未成熟果实果皮颜色：暗墨绿色
成熟果实果皮颜色：亮黄色
果颈：无
果基：平
果顶：钝圆
果实小果能否剥离：不可剥离
果眼外观：扁平或微凹
果眼深度：较浅
果眼排列方式：左旋或右旋
果瘤：多
果肉颜色：淡黄色
果实香味：无
果实风味：酸
果肉质地：脆/爽脆
果实外观综合评价：中
果实品质综合评价：差
果肉可溶性固形物含量：10.5%
果实成熟特性：中晚熟
成熟期的一致性：不一致
丰产性：低产

Black Pineapple

Code: ACCESSION000038
Name: Black Pineapple
Place of origin: unknown
Resource type: domestically collected variety
Main ways of consumption: fresh or processing
Material resource: SSCRI, CATAS collected from Ruili, Yunnan Province in 2008
Plant posture: spreading
Planting time: mid-February in 2021
Open heart stage: mid-March in 2022
Vegetative growth period: approx. 390 d
Initial flowering time: late March in 2022
Flowering period: approx. 22 d
Fruit maturation time: from mid-June to late July in 2022
Fruit developing period: approx. 100 d
Fruit shape: ellipsoid
Single fruit weight: 629 g
Longitudinal diameter: 11.8 cm
Transverse diameter: 9.5 cm
Fruit shape index: 1.2
Immature fruit peel color: dark blackish green
Mature fruit peel color: bright yellow
Fruit neck: absent
Fruit base: flat
Fruit top: bluntly round
Fruitlets adhesion situation: inseparable
Fruit eyes appearance: flat or slightly concave
Fruit eyes depth: relatively shallow
Fruit eyes arrangement: levorotation or dextrorotation
Fruit tumors: many
Flesh color: pale yellow
Fruit aroma: no scent
Fruit flavour: sour
Flesh texture: crisp/crunchy
Fruit appearance comprehensive evaluation: medium
Fruit quality comprehensive evaluation: poor
Soluble solids content in flesh: 10.5 %
Fruit ripening characteristics: medium-late mature
Maturity consistency: nonuniform
Productivity: low

菠萝种质资源图谱 （上册）
PINEAPPLE GERMPLASM RESOURCES MAP （Volume 1）

植株
Plant

现红
Open heart

花
Flower

花序
Inflorescence

第二章　国内菠萝种质资源
Chapter 2　Pineapple Germplasm Resources in China

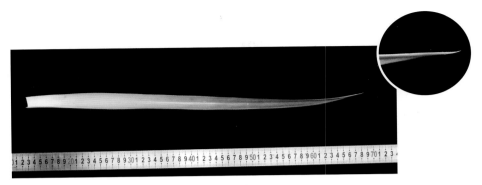

叶片
Leaf

带冠芽果
Fruit with crown bud

冠芽叶刺
Leaf spines in crown bud

果实
Fruit

果实纵切
Fruit longitudinal section

果实横切
Fruit transverse section

神湾菠萝

编号：ACCESSION000055
种质名称：神湾菠萝
原产地：新加坡
资源类型：国内收集
主要用途：鲜食或加工
种质来源地：2008年中国热带农业科学院南亚热带作物研究所从广东省广州市收集
植株姿态：开张
定植期：2021年2月下旬
现红期：2022年2月上旬
营养生长期：约345 d
初花期：2022年3月中旬
花开放时间：约22 d
成熟期：2022年6月上中旬
果实发育期：约85 d
果实形状：圆筒形
单果质量：453 g
纵径：10.9 cm
横径：9.0 cm
果形指数：1.2

未成熟果实果皮颜色：暗绿色
成熟果实果皮颜色：金黄/鲜黄色
果颈：无
果基：突起
果顶：平
果实小果能否剥离：可剥离
果眼外观：突起/隆起
果眼深度：深
果眼排列方式：左旋
果瘤：无
果肉颜色：黄色
果实香味：清香/微香
果实风味：酸甜
果肉质地：脆/爽脆
果实外观综合评价：优
果实品质综合评价：好
果肉可溶性固形物含量：13.4%
果实成熟特性：中熟
成熟期的一致性：基本一致
丰产性：低产

Shenwan Pineapple

Code: ACCESSION000055
Name: Shenwan Pineapple
Place of origin: Singapore
Resource type: domestically collected variety
Main ways of consumption: fresh or processing
Material resource: SSCRI, CATAS collected from Guangzhou, Guangdong Province in 2008
Plant posture: spreading
Planting time: late February in 2021
Open heart stage: early March in 2022
Vegetative growth period: approx. 345 d
Initial flowering time: mid-March in 2022
Flowering period: approx. 22 d
Fruit maturation time: early-to-mid July in 2022
Fruit developing period: approx. 85 d
Fruit shape: cylindrical
Single fruit weight: 453 g
Longitudinal diameter: 10.9 cm
Transverse diameter: 9.0 cm
Fruit shape index: 1.2
Immature fruit peel color: dark green
Mature fruit peel color: golden yellow/vivid yellow
Fruit neck: absent
Fruit base: protrusive
Fruit top: flat
Fruitlets adhesion situation: separable
Fruit eyes appearance: convex/embossed
Fruit eyes depth: deep
Fruit eyes arrangement: levorotation
Fruit tumors: absent
Flesh color: yellow
Fruit aroma: faint/slight scent
Fruit flavour: sour-sweet
Flesh texture: crisp/crunchy
Fruit appearance comprehensive evaluation: excellent
Fruit quality comprehensive evaluation: good
Soluble solids content in flesh: 13.4 %
Fruit ripening characteristics: medium mature
Maturity consistency: basically uniform
Productivity: low

菠萝种质资源图谱 （上册）
Pineapple Germplasm Resources Map (Volume 1)

植株
Plant

现红
Open heart

花
Flower

花序
Inflorescence

第二章 国内菠萝种质资源
Chapter 2　Pineapple Germplasm Resources in China

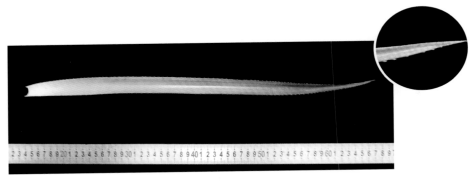

叶片
Leaf

带冠芽果
Fruit with crown bud

冠芽叶刺
Leaf spines in crown bud

果实
Fruit

果实纵切
Fruit longitudinal section

果实横切
Fruit transverse section

上海 2 号

编号：ACCESSION000106
种质名称：上海 2 号
原产地：不详
资源类型：国内收集
主要用途：鲜食或加工
种质来源地：2015 年中国热带农业科学院南亚热带作物研究所从海南省收集
植株姿态：开张
定植期：2021 年 2 月上旬
现红期：2022 年 2 月中旬
营养生长期：约 380 d
初花期：2022 年 3 月中旬
花开放时间：约 18 d
成熟期：2022 年 6 月下旬至 7 月中旬
果实发育期：约 110 d
果实形状：圆台形 / 方形
单果质量：991 g
纵径：13.3 cm
横径：10.9 cm
果形指数：1.2

未成熟果实果皮颜色：暗绿
成熟果实果皮颜色：亮黄 / 淡黄色
果颈：无
果基：平
果顶：平
果实小果能否剥离：不可剥离
果眼外观：扁平或微凹
果眼深度：浅
果眼排列方式：左旋
果瘤：无
果肉颜色：淡黄色
果实香味：芳香
果实风味：酸甜
果肉质地：滑
果实外观综合评价：中
果实品质综合评价：中
果肉可溶性固形物含量：15.4%
果实成熟特性：晚熟
成熟期的一致性：不一致
丰产性：一般

Shanghai No. 2

Code: ACCESSION000106
Name: Shanghai No. 2
Place of origin: unknown
Resource type: domestically collected variety
Main ways of consumption: fresh or processing
Material resource: SSCRI, CATAS collected from Hainan Province in 2015
Plant posture: spreading
Planting time: early February in 2021
Open heart stage: mid-February in 2022
Vegetative growth period: approx. 380 d
Initial flowering time: mid-March in 2022
Flowering period: approx. 18 d
Fruit maturation time: from late June to mid-July in 2022
Fruit developing period: approx. 110 d
Fruit shape: truncated cone/square
Single fruit weight: 991 g
Longitudinal diameter: 13.3 cm
Transverse diameter: 10.9 cm
Fruit shape index: 1.2
Immature fruit peel color: dark green
Mature fruit peel color: bright yellow/light yellow
Fruit neck: absent
Fruit base: flat
Fruit top: flat
Fruitlets adhesion situation: inseparable
Fruit eyes appearance: flat or slightly concave
Fruit eyes depth: shallow
Fruit eyes arrangement: levorotation
Fruit tumors: absent
Flesh color: pale yellow
Fruit aroma: sweet fragrance
Fruit flavour: sour-sweet
Flesh texture: smooth
Fruit appearance comprehensive evaluation: medium
Fruit quality comprehensive evaluation: medium
Soluble solids content in flesh: 15.4 %
Fruit ripening characteristics: late mature
Maturity consistency: nonuniform
Productivity: normal level

菠萝种质资源图谱 （上册）
Pineapple Germplasm Resources Map (Volume 1)

植株
Plant

现红
Open heart

花
Flower

花序
Inflorescence

第二章　国内菠萝种质资源
Chapter 2　Pineapple Germplasm Resources in China

叶片
Leaf

带冠芽果
Fruit with crown bud

冠芽叶刺
Leaf spines in crown bud

果实
Fruit

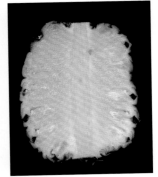

果实纵切
Fruit longitudinal section

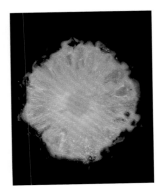

果实横切
Fruit transverse section

红皮菠萝

编号：ACCESSION000124
种质名称：红皮菠萝，又名红果
原产地：不详
资源类型：国内收集
主要用途：观赏
种质来源地：2016年中国热带农业科学院南亚热带作物研究所从海南省收集
植株姿态：开张
定植期：2021年2月上旬
现红期：2022年3月中旬
营养生长期：约400 d
初花期：2022年4月上旬
花开放时间：约17 d
成熟期：2022年6月下旬至7月下旬
果实发育期：约105 d
果实形状：长圆筒形
单果质量：650 g
纵径：12.7 cm
横径：9.4 cm
果形指数：1.4

未成熟果实果皮颜色：橙红色
成熟果实果皮颜色：橙红略显黄白色
果颈：无
果基：突起
果顶：平
果实小果能否剥离：不可剥离
果眼外观：扁平或微凹
果眼深度：浅
果眼排列方式：左旋或右旋
果瘤：多
果肉颜色：奶油色
果实香味：无
果实风味：酸
果肉质地：脆/爽脆
果实外观综合评价：优
果实品质综合评价：差
果肉可溶性固形物含量：8.1%
果实成熟特性：中晚熟
成熟期的一致性：不一致
丰产性：低产

Red skin Pineapple

Code: ACCESSION000124
Name: Red Skin Pineapple, also known as Red Fruit
Place of origin: unknown
Resource type: domestically collected variety
Main ways of consumption: ornamentation
Material resource: SSCRI, CATAS collected from Hainan in 2016
Plant posture: spreading
Planting time: early February in 2021
Open heart stage: mid-March in 2022
Vegetative growth period: approx. 400 d
Initial flowering time: early April in 2022
Flowering period: approx. 17 d
Fruit maturation time: from late June to late July in 2022
Fruit developing period: approx. 105 d
Fruit shape: long cylindrical
Single fruit weight: 650 g
Longitudinal diameter: 12.7 cm
Transverse diameter: 9.4 cm
Fruit shape index: 1.4
Immature fruit peel color: orange red
Mature fruit peel color: orange red with yellowish-white
Fruit neck: absent
Fruit base: protrusive
Fruit top: flat
Fruitlets adhesion situation: inseparable
Fruit eyes appearance: flat or slightly concave
Fruit eyes depth: shallow
Fruit eyes arrangement: levorotation or dextrorotation
Fruit tumors: many
Flesh color: cream
Fruit aroma: no scent
Fruit flavour: sour
Flesh texture: crisp/crunchy
Fruit appearance comprehensive evaluation: excellent
Fruit quality comprehensive evaluation: poor
Soluble solids content in flesh: 8.1 %
Fruit ripening characteristics: medium-late mature
Maturity consistency: nonuniform
Productivity: low

菠萝种质资源图谱（上册）
Pineapple Germplasm Resources Map *(Volume 1)*

植株
Plant

现红
Open heart

花
Flower

花序
Inflorescence

第二章　国内菠萝种质资源
Chapter 2　Pineapple Germplasm Resources in China

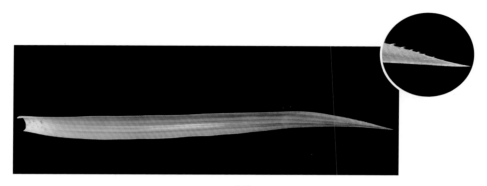

叶片
Leaf

带冠芽果
Fruit with crown bud

冠芽叶刺
Leaf spines in crown bud

果实
Fruit

果实纵切
Fruit longitudinal section

果实横切
Fruit transverse section

第三章
国外菠萝种质资源

―――――

Chapter 3
Pineapple Germplasm
Resources from Abroad

MD-2

编号：ACCESSION000012
种质名称：MD-2，又名 73-114、金菠萝
原产地：美国
资源类型：引进品种
主要用途：鲜食或加工
种质来源地：2006 年中国热带农业科学院南亚热带作物研究所从澳大利亚引进
植株姿态：开张
定植期：2021 年 2 月中旬
现红期：2022 年 2 月中旬
营养生长期：约 360 d
初花期：2022 年 3 月下旬
花开放时间：约 22 d
成熟期：2022 年 6 月下旬至 7 月上旬
果实发育期：约 95 d
果实形状：圆台形 / 方形
单果质量：905 g
纵径：11.8 cm
横径：10.9 cm
果形指数：1.1
未成熟果实果皮颜色：暗绿色
成熟果实果皮颜色：暗黄 / 深黄色
果颈：无
果基：平
果顶：平
果实小果能否剥离：不可剥离
果眼外观：扁平或微凹
果眼深度：较浅
果眼排列方式：左旋、右旋或其他
果瘤：无
果肉颜色：金黄色
果实香味：清香 / 微香
果实风味：甜酸
果肉质地：粗糙
果实外观综合评价：中
果实品质综合评价：好
果肉可溶性固形物含量：17.1 %
果实成熟特性：中熟
成熟期的一致性：不一致
丰产性：一般

MD-2

Code: ACCESSION000012

Name: MD-2, also known as 73-114 or Gold Pineapple

Place of origin: America

Resource type: introduced variety

Main ways of consumption: fresh or processing

Initially introduction place: SSCRI, CATAS introduced from Australia in 2006

Plant posture: spreading

Planting time: mid-February in 2021

Open heart stage: mid- February in 2022

Vegetative growth period: approx. 360 d

Initial flowering time: late March in 2022

Flowering period: approx. 22 d

Fruit maturation time: from late June to early July in 2022

Fruit developing period: approx. 95 d

Fruit shape: truncated cone/square

Single fruit weight: 905 g

Longitudinal diameter: 11.8 cm

Transverse diameter: 10.9 cm

Fruit shape index: 1.1

Immature fruit peel color: dark green

Mature fruit peel color: dark yellow/deep yellow

Fruit neck: absent

Fruit base: flat

Fruit top: flat

Fruitlets adhesion situation: inseparable

Fruit eyes appearance: flat or slightly concave

Fruit eyes depth: relatively shallow

Fruit eyes arrangement: levorotation, dextrorotation or other

Fruit tumors: absent

Flesh color: golden yellow

Fruit aroma: faint/slight scent

Fruit flavour: sweet-sour

Flesh texture: coarse

Fruit appearance comprehensive evaluation: medium

Fruit quality comprehensive evaluation: good

Soluble solids content in flesh: 17.1 %

Fruit ripening characteristics: medium mature

Maturity consistency: nonuniform

Productivity: normal level

植株
Plant

现红
Open heart

花
Flower

花序
Inflorescence

第三章 国外菠萝种质资源
Chapter 3 Pineapple Germplasm Resources from Abroad

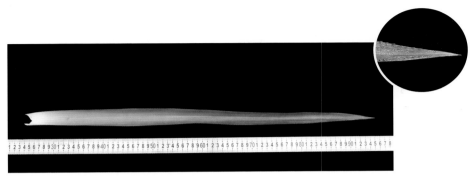

叶片
Leaf

带冠芽果
Fruit with crown bud

冠芽叶刺
Leaf spines in crown bud

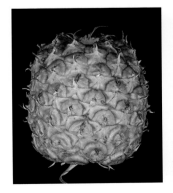

果实
Fruit

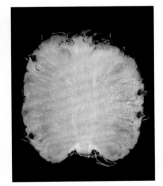

果实纵切
Fruit longitudinal section

果实横切
Fruit transverse section

金冬蜜

编号：ACCESSION000009
种质名称：金冬蜜
原产地：澳大利亚
资源类型：引进品种
主要用途：鲜食或加工
种质来源地：2006年中国热带农业科学院南亚热带作物研究所从澳大利亚引进
植株姿态：开张
定植期：2021年2月上旬
现红期：2022年2月中旬
营养生长期：约380 d
初花期：2022年3月中旬
花开放时间：约18 d
成熟期：2022年6月中旬
果实发育期：约90 d
果实形状：圆筒形
单果质量：885 g
纵径：12.2 cm
横径：10.8 cm
果形指数：1.1

未成熟果实果皮颜色：暗绿色
成熟果实果皮颜色：亮黄/淡黄色
果颈：无
果基：平
果顶：平
果实小果能否剥离：可剥离
果眼外观：微隆起
果眼深度：较深
果眼排列方式：左旋或右旋
果瘤：无
果肉颜色：金黄色
果实香味：清香/微香
果实风味：浓甜
果肉质地：脆/爽脆
果实外观综合评价：优
果实品质综合评价：优
果肉可溶性固形物含量：23.9%
果实成熟特性：中熟
成熟期的一致性：一致
丰产性：一般

Golden Winter Sweet

Code: ACCESSION00009
Name: Golden Winter Sweet
Place of origin: Australia
Resource type: introduced variety
Main ways of consumption: fresh or processing
Initially introduction place: SSCRI, CATAS introduced from Australia in 2006
Plant posture: spreading
Planting time: early February in 2021
Open heart stage: mid- February in 2022
Vegetative growth period: approx. 380 d
Initial flowering time: mid-March in 2022
Flowering period: approx. 18 d
Fruit maturation time: mid-June in 2022
Fruit developing period: approx. 90 d
Fruit shape: cylindrical
Single fruit weight: 885 g
Longitudinal diameter: 12.2 cm
Transverse diameter: 10.8 cm
Fruit shape index: 1.1
Immature fruit peel color: dark green
Mature fruit peel color: bright yellow/ light yellow
Fruit neck: absent
Fruit base: flat
Fruit top: flat
Fruitlets adhesion situation: separable
Fruit eyes appearance: slightly embossed
Fruit eyes depth: relatively deep
Fruit eyes arrangement: levorotation or dextrorotation
Fruit tumors: absent
Flesh color: golden yellow
Fruit aroma: faint/slight scent
Fruit flavour: strong sweet
Flesh texture: crisp/crunchy
Fruit appearance comprehensive evaluation: excellent
Fruit quality comprehensive evaluation: excellent
Soluble solids content in flesh: 23.9 %
Fruit ripening characteristics: medium mature
Maturity consistency: uniform
Productivity: normal level

菠萝种质资源图谱 （上册）
Pineapple Germplasm Resources Map (Volume 1)

植株
Plant

现红
Open heart

花
Flower

花序
Inflorescence

第三章　国外菠萝种质资源
Chapter 3　Pineapple Germplasm Resources from Abroad

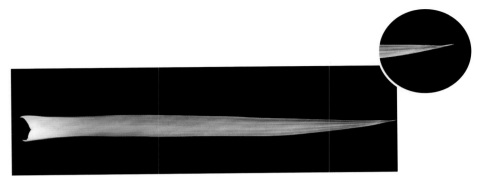

叶片
Leaf

带冠芽果
Fruit with crown bud

冠芽叶刺
Leaf spines in crown bud

果实
Fruit

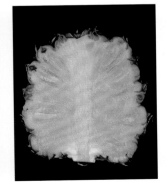

果实纵切
Fruit longitudinal section

果实横切
Fruit transverse section

鲜优菠萝

编号：ACCESSION000014
种质名称：鲜优菠萝
原产地：澳大利亚
资源类型：引进品种
主要用途：鲜食或加工
种质来源地：2006年中国热带农业科学院南亚热带作物研究所从澳大利亚引进
植株姿态：开张
定植期：2021年2月下旬
现红期：2022年2月中旬
营养生长期：约360 d
初花期：2022年3月中旬
花开放时间：约24 d
成熟期：2022年6月中旬
果实发育期：约90 d
果实形状：（近）圆球形
单果质量：518 g
纵径：10.0 cm
横径：9.7 cm
果形指数：1.0

未成熟果实果皮颜色：银绿色
成熟果实果皮颜色：亮黄/淡黄色
果颈：无
果基：弧形
果顶：平
果实小果能否剥离：可剥离
果眼外观：突起/隆起
果眼深度：深
果眼排列方式：左旋或右旋
果瘤：无
果肉颜色：金黄色
果实香味：清香/微香
果实风味：甜酸
果肉质地：滑
果实外观综合评价：好
果实品质综合评价：优
果肉可溶性固形物含量：20.4%
果实成熟特性：中熟
成熟期的一致性：基本一致
丰产性：低产

Fresh Premium Pineapple

Code: ACCESSION000014
Name: Fresh Premium Pineapple
Place of origin: Australia
Resource type: introduced variety
Main ways of consumption: fresh or processing
Initially introduction place: SSCRI, CATAS introduced from Australia in 2006
Plant posture: spreading
Planting time: late February in 2021
Open heart stage: mid-February in 2022
Vegetative growth period: approx. 360 d
Initial flowering time: mid-March in 2022
Flowering period: approx. 24 d
Fruit maturation time: mid-June in 2022
Fruit developing period: approx. 90 d
Fruit shape: (near) spherical
Single fruit weight: 518 g
Longitudinal diameter: 10.0 cm
Transverse diameter: 9.7 cm
Fruit shape index: 1.0
Immature fruit peel color: silver green
Mature fruit peel color: bright yellow/ light yellow
Fruit neck: absent
Fruit base: curved
Fruit top: flat
Fruitlets adhesion situation: separable
Fruit eyes appearance: convex/ embossed
Fruit eyes depth: deep
Fruit eyes arrangement: levorotation or dextrorotation
Fruit tumors: absent
Flesh color: golden yellow
Fruit aroma: faint/slight scent
Fruit flavour: sweet-sour
Flesh texture: smooth
Fruit appearance comprehensive evaluation: good
Fruit quality comprehensive evaluation: excellent
Soluble solids content in flesh: 20.4 %
Fruit ripening characteristics: medium mature
Maturity consistency: basically uniform
Productivity: low

菠萝种质资源图谱（上册）
Pineapple Germplasm Resources Map (Volume 1)

植株
Plant

现红
Open heart

花
Flower

花序
Inflorescence

第三章　国外菠萝种质资源
Chapter 3　Pineapple Germplasm Resources from Abroad

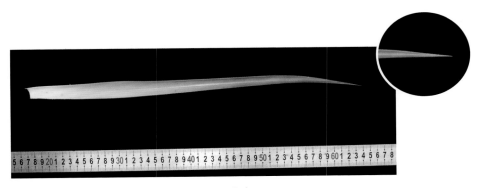

叶片
Leaf

带冠芽果
Fruit with crown bud

冠芽叶刺
Leaf spines in crown bud

果实
Fruit

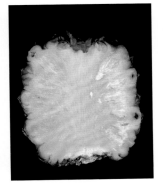

果实纵切
Fruit longitudinal section

果实横切
Fruit transverse section

马若奇

编号：ACCESSION000013

种质名称：马若奇

原产地：澳大利亚

资源类型：引进品种

主要用途：鲜食或加工

种质来源地：2006年中国热带农业科学院南亚热带作物研究所从澳大利亚引进

植株姿态：开张

定植期：2021年2月下旬

现红期：2022年3月中旬

营养生长期：约390 d

初花期：2022年4月上旬

花开放时间：约19 d

成熟期：2022年7月下旬至8月下旬

果实发育期：约125 d

果实形状：圆筒形

单果质量：424 g

纵径：10.1 cm

横径：8.2 cm

果形指数：1.2

未成熟果实果皮颜色：暗墨绿色

成熟果实果皮颜色：暗黄/深黄色

果颈：无

果基：平

果顶：平

果实小果能否剥离：可剥离

果眼外观：扁平或微凹

果眼深度：较深

果眼排列方式：左旋或右旋

果瘤：无

果肉颜色：淡黄色

果实香味：无

果实风味：甜酸

果肉质地：粗糙

果实外观综合评价：优

果实品质综合评价：中

果肉可溶性固形物含量：16.7%

果实成熟特性：晚熟

成熟期的一致性：不一致

丰产性：低产

Maroochy

Code: ACCESSION000013
Name: Maroochy
Place of origin: Australia
Resource type: introduced variety
Main ways of consumption: fresh or processing
Initially introduction place: SSCRI, CATAS introduced from Australia in 2006
Plant posture: spreading
Planting time: late February in 2021
Open heart stage: mid-March in 2022
Vegetative growth period: approx. 390 d
Initial flowering time: early April in 2022
Flowering period: approx. 19 d
Fruit maturation time: from late July to late August in 2022
Fruit developing period: approx. 125 d
Fruit shape: cylindrical
Single fruit weight: 424 g
Longitudinal diameter: 10.1cm
Transverse diameter: 8.2 cm
Fruit shape index: 1.2
Immature fruit peel color: dark blackish green
Mature fruit peel color: dark yellow/deep yellow
Fruit neck: absent
Fruit base: flat
Fruit top: flat
Fruitlets adhesion situation: separable
Fruit eyes appearance: flat or slightly concave
Fruit eyes depth: relatively deep
Fruit eyes arrangement: levorotation or dextrorotation
Fruit tumors: absent
Flesh color: pale yellow
Fruit aroma: no scent
Fruit flavour: sweet-sour
Flesh texture: coarse
Fruit appearance comprehensive evaluation: excellent
Fruit quality comprehensive evaluation: medium
Soluble solids content in flesh: 16.7 %
Fruit ripening characteristics: late mature
Maturity consistency: nonuniform
Productivity: low

菠萝种质资源图谱 （上册）
Pineapple Germplasm Resources Map (Volume 1)

植株
Plant

现红
Open heart

花
Flower

花序
Inflorescence

第三章　国外菠萝种质资源
Chapter 3　Pineapple Germplasm Resources from Abroad

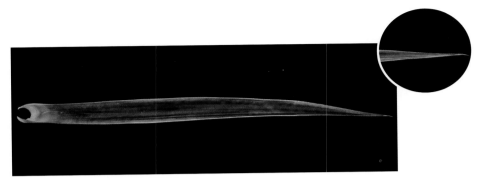

叶片
Leaf

带冠芽果
Fruit with crown bud

冠芽叶刺
Leaf spines in crown bud

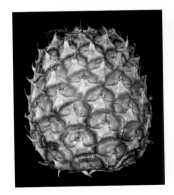

果实
Fruit

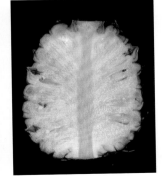

果实纵切
Fruit longitudinal section

果实横切
Fruit transverse section

麦格雷戈

编号：BL00127
种质名称：麦格雷戈
原产地：南非
资源类型：引进品种
主要用途：鲜食或加工
种质来源地：2006年中国热带农业科学院南亚热带作物研究所从澳大利亚引进
植株姿态：开张
定植期：2021年2月下旬
现红期：2022年2月上旬
营养生长期：约345 d
初花期：2022年3月中旬
花开放时间：约21 d
成熟期：2022年6月中下旬
果实发育期：约95 d
果实形状：圆筒形
单果质量：536 g
纵径：10.8 cm
横径：9.4 cm
果形指数：1.2

未成熟果实果皮颜色：银绿色
成熟果实果皮颜色：金黄/鲜黄色
果颈：无
果基：平
果顶：平
果实小果能否剥离：可剥离
果眼外观：微隆起
果眼深度：较深
果眼排列方式：左旋
果瘤：无
果肉颜色：金黄色
果实香味：清香/微香
果实风味：酸甜
果肉质地：滑
果实外观综合评价：优
果实品质综合评价：好
果肉可溶性固形物含量：16.9%
果实成熟特性：晚熟
成熟期的一致性：基本一致
丰产性：低产

MacGregor

Code: BL00127
Name: MacGregor
Place of origin: Africa
Resource type: introduced variety
Main ways of consumption: fresh or processing
Initially introduction place: SSCRI, CATAS introduced from Australia in 2006
Plant posture: spreading
Planting time: late February in 2021
Open heart stage: early February in 2022
Vegetative growth period: approx. 345 d
Initial flowering time: mid-March in 2022
Flowering period: approx. 21 d
Fruit maturation time: mid-to-late June in 2022
Fruit developing period: approx. 95 d
Fruit shape: cylindrical
Single fruit weight: 536 g
Longitudinal diameter: 10.8 cm
Transverse diameter: 9.4 cm
Fruit shape index: 1.2
Immature fruit peel color: silver green
Mature fruit peel color: golden yellow/vivid yellow
Fruit neck: absent
Fruit base: flat
Fruit top: flat
Fruitlets adhesion situation: separable
Fruit eyes appearance: slightly embossed
Fruit eyes depth: relatively deep
Fruit eyes arrangement: levorotation
Fruit tumors: absent
Flesh color: golden yellow
Fruit aroma: faint/slight scent
Fruit flavour: sour-sweet
Flesh texture: smooth
Fruit appearance comprehensive evaluation: excellent
Fruit quality comprehensive evaluation: good
Soluble solids content in flesh: 16.9 %
Fruit ripening characteristics: late mature
Maturity consistency: basically uniform
Productivity: low

菠萝种质资源图谱 （上册）
Pineapple Germplasm Resources Map *(Volume 1)*

植株
Plant

现红
Open heart

花
Flower

花序
Inflorescence

第三章 国外菠萝种质资源
Chapter 3 Pineapple Germplasm Resources from Abroad

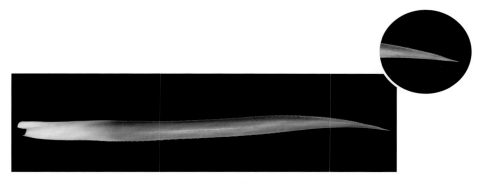

叶片
Leaf

带冠芽果
Fruit with crown bud

冠芽叶刺
Leaf spines in crown bud

果实
Fruit

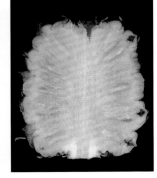

果实纵切
Fruit longitudinal section

果实横切
Fruit transverse section

碧武里 1 号

编号：ACCESSION000022
种质名称：碧武里 1 号
原产地：泰国
资源类型：引进品种
主要用途：鲜食或加工
种质来源地：2006 年中国热带农业科学院南亚热带作物研究所从泰国引进
植株姿态：开张
定植期：2021 年 2 月上旬
现红期：2022 年 2 月中旬
营养生长期：约 380 d
初花期：2022 年 3 月下旬
花开放时间：约 13 d
成熟期：2022 年 5 月下旬
果实发育期：约 60 d
果实形状：长圆筒形
单果质量：880 g
纵径：15.2 cm
横径：10.0 cm
果形指数：1.5

未成熟果实果皮颜色：暗绿
成熟果实果皮颜色：亮黄
果颈：无
果基：弧形
果顶：钝圆
果实小果能否剥离：可剥离
果眼外观：突起 / 隆起
果眼深度：深
果眼排列方式：左旋或右旋
果瘤：少
果肉颜色：金黄色
果实香味：清香 / 微香
果实风味：清甜
果肉质地：粗糙
果实外观综合评价：中
果实品质综合评价：好
果肉可溶性固形物含量：19.2%
果实成熟特性：特早熟
成熟期的一致性：一致
丰产性：一般

Phetchaburi No. 1

Code: ACCESSION000022
Name: Phetchaburi No. 1
Place of origin: Thailand
Resource type: introduced variety
Main ways of consumption: fresh or processing
Initially introduction place: SSCRI, CATAS introduced from Thailand in 2006
Plant posture: spreading
Planting time: early February in 2021
Open heart stage: mid-February in 2022
Vegetative growth period: approx. 380 d
Initial flowering time: late March in 2022
Flowering period: approx. 13 d
Fruit maturation time: late May in 2022
Fruit developing period: approx. 60 d
Fruit shape: long cylindrical
Single fruit weight: 880 g
Longitudinal diameter: 15.2 cm
Transverse diameter: 10.0 cm
Fruit shape index: 1.5
Immature fruit peel color: dark green
Mature fruit peel color: bright yellow
Fruit neck: absent
Fruit base: curved
Fruit top: bluntly round
Fruitlets adhesion situation: separable
Fruit eyes appearance: convex/embossed
Fruit eyes depth: deep
Fruit eyes arrangement: levorotation or dextrorotation
Fruit tumors: few
Flesh color: golden yellow
Fruit aroma: faint/slight scent
Fruit flavour: mildly sweet
Flesh texture: coarse
Fruit appearance comprehensive evaluation: medium
Fruit quality comprehensive evaluation: good
Soluble solids content in flesh: 19.2 %
Fruit ripening characteristics: extremely early mature
Maturity consistency: uniform
Productivity: normal level

菠萝种质资源图谱 （上册）
Pineapple Germplasm Resources Map (Volume 1)

植株
Plant

现红
Open heart

花
Flower

花序
Inflorescence

第三章　国外菠萝种质资源
Chapter 3　Pineapple Germplasm Resources from Abroad

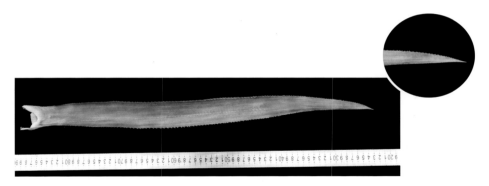

叶片
Leaf

带冠芽果
Fruit with crown bud

冠芽叶刺
Leaf spines in crown bud

果实
Fruit

果实纵切
Fruit longitudinal section

果实横切
Fruit transverse section

碧武里 2 号

编号：ACCESSION000023
种质名称：碧武里 2 号
原产地：泰国
资源类型：引进品种
主要用途：鲜食或加工
种质来源地：2006 年中国热带农业科学院南亚热带作物研究所从泰国引进
植株姿态：开张
定植期：2021 年 2 月上旬
现红期：2022 年 3 月中旬
营养生长期：约 400 d
初花期：2022 年 3 月下旬
花开放时间：约 30 d
成熟期：2022 年 6 月下旬至 8 月下旬
果实发育期：约 120 d
果实形状：圆筒形
单果质量：1109 g
纵径：15.0 cm
横径：11.1 cm
果形指数：1.4

未成熟果实果皮颜色：暗绿色
成熟果实果皮颜色：淡黄色
果颈：无
果基：平
果顶：平
果实小果能否剥离：不可剥离
果眼外观：扁平或微凹
果眼深度：较浅
果眼排列方式：左旋或右旋
果瘤：无
果肉颜色：黄色
果实香味：清香 / 微香
果实风味：酸
果肉质地：脆 / 爽脆
果实外观综合评价：中
果实品质综合评价：中
果肉可溶性固形物含量：14.3%
果实成熟特性：中晚熟
成熟期的一致性：不一致
丰产性：丰产

Phetchaburi No. 2

Code: ACCESSION000023
Name: Phetchaburi No. 2
Place of origin: Thailand
Resource type: introduced variety
Main ways of consumption: fresh or processing
Initially introduction place: SSCRI, CATAS introduced from Thailand in 2006
Plant posture: spreading
Planting time: early February in 2021
Open heart stage: mid-March in 2022
Vegetative growth period: approx. 400 d
Initial flowering time: late March in 2022
Flowering period: approx. 30 d
Fruit maturation time: from late June to late August in 2022
Fruit developing period: approx. 120 d
Fruit shape: cylindrical
Single fruit weight: 1109 g
Longitudinal diameter: 15.0 cm
Transverse diameter: 11.1 cm
Fruit shape index: 1.4
Immature fruit peel color: dark green
Mature fruit peel color: light yellow
Fruit neck: absent
Fruit base: flat
Fruit top: flat
Fruitlets adhesion situation: inseparable
Fruit eyes appearance: flat or slightly concave
Fruit eyes depth: relatively shallow
Fruit eyes arrangement: levorotation or dextrorotation
Fruit tumors: absent
Flesh color: yellow
Fruit aroma: faint/slight scent
Fruit flavour: sour
Flesh texture: crisp/crunchy
Fruit appearance comprehensive evaluation: medium
Fruit quality comprehensive evaluation: medium
Soluble solids content in flesh: 14.3 %
Fruit ripening characteristics: medium-late mature
Maturity consistency: nonuniform
Productivity: productive

菠萝种质资源图谱 （上册）
Pineapple Germplasm Resources Map (Volume 1)

植株
Plant

现红
Open heart

花
Flower

花序
Inflorescence

第三章 国外菠萝种质资源
Chapter 3 Pineapple Germplasm Resources from Abroad

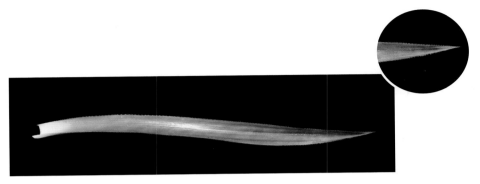

叶片
Leaf

带冠芽果
Fruit with crown bud

冠芽叶刺
Leaf spines in crown bud

果实
Fruit

果实纵切
Fruit longitudinal section

果实横切
Fruit transverse section

普吉

编号：ACCESSION000027
种质名称：普吉
原产地：泰国
资源类型：引进品种
主要用途：鲜食或加工
种质来源地：2007年中国热带农业科学院南亚热带作物研究所从泰国引进
植株姿态：开张
定植期：2021年2月上旬
现红期：2022年2月上旬
营养生长期：约365 d
初花期：2022年3月下旬
花开放时间：约20 d
成熟期：2022年6月上中旬
果实发育期：约75 d
果实形状：圆筒形
单果质量：992 g
纵径：14.8 cm
横径：11.1 cm
果形指数：1.3

未成熟果实果皮颜色：暗绿色
成熟果实果皮颜色：亮黄/淡黄色
果颈：无
果基：突起
果顶：钝圆
果实小果能否剥离：不可剥离
果眼外观：突起/隆起
果眼深度：深
果眼排列方式：左旋或右旋
果瘤：少
果肉颜色：淡黄色
果实香味：清香/微香
果实风味：清甜
果肉质地：脆/爽脆
果实外观综合评价：中
果实品质综合评价：中
果肉可溶性固形物含量：13.8%
果实成熟特性：中熟
成熟期的一致性：不一致
丰产性：一般

Phuket

Code: ACCESSION000027
Name: Phuket
Place of origin: Thailand
Resource type: introduced variety
Main ways of consumption: fresh or processing
Initially introduction place: SSCRI, CATAS introduced from Thailand in 2007
Plant posture: spreading
Planting time: early February in 2021
Open heart stage: early February in 2022
Vegetative growth period: approx. 365 d
Initial flowering time: late March in 2022
Flowering period: approx. 20 d
Fruit maturation time: early-to-mid June in 2022
Fruit developing period: approx. 75 d
Fruit shape: cylindrical
Single fruit weight: 992 g
Longitudinal diameter: 14.8 cm
Transverse diameter: 11.1 cm
Fruit shape index: 1.3
Immature fruit peel color: dark green
Mature fruit peel color: bright yellow/light yellow
Fruit neck: absent
Fruit base: protrusive
Fruit top: bluntly round
Fruitlets adhesion situation: inseparable
Fruit eyes appearance: convex/embossed
Fruit eyes depth: deep
Fruit eyes arrangement: levorotation or dextrorotation
Fruit tumors: few
Flesh color: pale yellow
Fruit aroma: faint/slight scent
Fruit flavour: mildly sweet
Flesh texture: crisp/crunchy
Fruit appearance comprehensive evaluation: medium
Fruit quality comprehensive evaluation: medium
Soluble solids content in flesh: 13.8 %
Fruit ripening characteristics: medium mature
Maturity consistency: nonuniform
Productivity: normal level

植株
Plant

现红
Open heart

花序
Inflorescence

花
Flower

第三章　国外菠萝种质资源
Chapter 3　Pineapple Germplasm Resources from Abroad

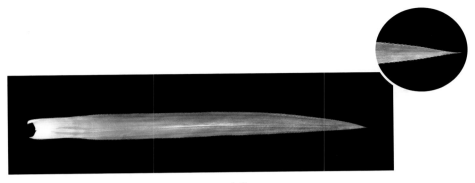

叶片
Leaf

带冠芽果

Fruit with crown bud

冠芽叶刺

Leaf spines in crown bud

果实
Fruit

果实纵切
Fruit longitudinal section

果实横切
Fruit transverse section

· 97 ·

新普吉

编号：ACCESSION000026
种质名称：新普吉
原产地：泰国
资源类型：引进品种
主要用途：鲜食或加工
种质来源地：2007 年中国热带农业科学院南亚热带作物研究所从泰国引进
植株姿态：开张
定植期：2021 年 2 月下旬
现红期：2022 年 2 月中旬
营养生长期：约 360 d
初花期：2022 年 3 月中旬
花开放时间：约 20 d
成熟期：2022 年 6 月上中旬
果实发育期：约 85 d
果实形状：圆筒形
单果质量：602 g
纵径：12.5 cm
横径：9.5 cm
果形指数：1.3

未成熟果实果皮颜色：暗墨绿色
成熟果实果皮颜色：金黄/鲜黄色
果颈：无
果基：弧形
果顶：钝圆
果实小果能否剥离：可剥离
果眼外观：突起/隆起
果眼深度：深
果眼排列方式：左旋或右旋
果瘤：无
果肉颜色：淡黄色
果实香味：清香/微香
果实风味：清甜
果肉质地：脆/爽脆
果实外观综合评价：优
果实品质综合评价：好
果肉可溶性固形物含量：17.3%
果实成熟特性：中熟
成熟期的一致性：基本一致
丰产性：低产

Chapter 3 Pineapple Germplasm Resources from Abroad

New Phuket

Code: ACCESSION000026
Name: New Phuket
Place of origin: Thailand
Resource type: introduced variety
Main ways of consumption: fresh or processing
Initially introduction place: SSCRI, CATAS introduced from Thailand in 2007
Plant posture: spreading
Planting time: late February in 2021
Open heart stage: mid-February in 2022
Vegetative growth period: approx. 360 d
Initial flowering time: mid-March in 2022
Flowering period: approx. 20 d
Fruit maturation time: early-to-mid June in 2022
Fruit developing period: approx. 85 d
Fruit shape: cylindrical
Single fruit weight: 602 g
Longitudinal diameter: 12.5 cm
Transverse diameter: 9.5 cm
Fruit shape index: 1.3
Immature fruit peel color: dark blackish green
Mature fruit peel color: golden yellow/vivid yellow
Fruit neck: absent
Fruit base: curved
Fruit top: bluntly round
Fruitlets adhesion situation: separable
Fruit eyes appearance: convex/embossed
Fruit eyes depth: deep
Fruit eyes arrangement: levorotation, dextrorotation
Fruit tumors: absent
Flesh color: pale yellow
Fruit aroma: faint/slight scent
Fruit flavour: mildly sweet
Flesh texture: crisp/crunchy
Fruit appearance comprehensive evaluation: excellent
Fruit quality comprehensive evaluation: good
Soluble solids content in flesh: 17.3 %
Fruit ripening characteristics: medium mature
Maturity consistency: basically uniform
Productivity: low

菠萝种质资源图谱 （上册）
Pineapple Germplasm Resources Map (Volume 1)

植株
Plant

现红
Open heart

花
Flower

花序
Inflorescence

第三章 国外菠萝种质资源
Chapter 3 Pineapple Germplasm Resources from Abroad

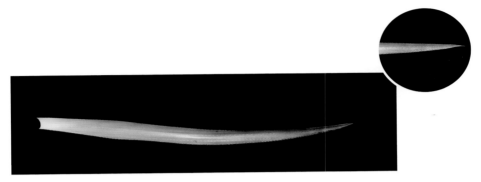

叶片
Leaf

带冠芽果
Fruit with crown bud

冠芽叶刺
Leaf spines in crown bud

果实
Fruit

果实纵切
Fruit longitudinal section

果实横切
Fruit transverse section

Tradsrithong

编号：ACCESSION000028
种质名称：Tradsrithong
原产地：泰国
资源类型：引进品种
主要用途：鲜食或加工
种质来源地：2007年中国热带农业科学院南亚热带作物研究所从泰国引进
植株姿态：开张
定植期：2021年2月下旬
现红期：2022年2月中旬
营养生长期：约350 d
初花期：2022年3月下旬
花开放时间：约19 d
成熟期：2022年6月上中旬
果实发育期：约75 d
果实形状：长圆柱形
单果质量：581 g
纵径：12.3 cm
横径：9.0 cm
果形指数：1.4

未成熟果实果皮颜色：暗绿色
成熟果实果皮颜色：亮黄/淡黄色
果颈：无
果基：平
果顶：平
果实小果能否剥离：可剥离
果眼外观：突起/隆起
果眼深度：深
果眼排列方式：左旋或右旋
果瘤：无
果肉颜色：淡黄色
果实香味：无
果实风味：酸甜
果肉质地：滑
果实外观综合评价：优
果实品质综合评价：好
果肉可溶性固形物含量：16.3%
果实成熟特性：早熟
成熟期的一致性：基本一致
丰产性：低产

Tradsrithong

Code: ACCESSION000028
Name: Tradsrithong
Place of origin: Thailand
Resource type: introduced variety
Main ways of consumption: fresh or processing
Initially introduction place: SSCRI, CATAS introduced from Thailand in 2007
Plant posture: spreading
Planting time: late February in 2021
Open heart stage: mid-February in 2022
Vegetative growth period: approx. 350 d
Initial flowering time: late March in 2022
Flowering period: approx. 19 d
Fruit maturation time: early-to-mid June in 2022
Fruit developing period: approx. 75 d
Fruit shape: long cylindrical
Single fruit weight: 581 g
Longitudinal diameter: 12.3 cm
Transverse diameter: 9.0 cm
Fruit shape index: 1.4
Immature fruit peel color: dark green
Mature fruit peel color: bright yellow/ light yellow
Fruit neck: absent
Fruit base: flat
Fruit top: flat
Fruitlets adhesion situation: separable
Fruit eyes appearance: convex/ embossed
Fruit eyes depth: deep
Fruit eyes arrangement: levorotation or dextrorotation
Fruit tumors: absent
Flesh color: pale yellow
Fruit aroma: no scent
Fruit flavour: sour-sweet
Flesh texture: smooth
Fruit appearance comprehensive evaluation: excellent
Fruit quality comprehensive evaluation: good
Soluble solids content in flesh: 16.3 %
Fruit ripening characteristics: early mature
Maturity consistency: basically uniform
Productivity: low

菠萝种质资源图谱（上册）
Pineapple Germplasm Resources Map (Volume 1)

植株
Plant

现红
Open heart

花
Flower

花序
Inflorescence

第三章 国外菠萝种质资源
Chapter 3　Pineapple Germplasm Resources from Abroad

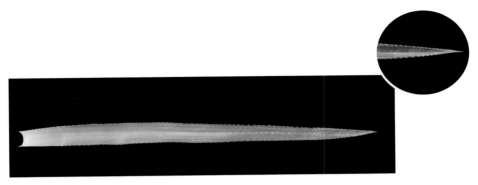

叶片
Leaf

带冠芽果
Fruit with crown bud

冠芽叶刺
Leaf spines in crown bud

果实
Fruit

果实纵切
Fruit longitudinal section

果实横切
Fruit transverse section

无刺卡因（巴冲）

编号：ACCESSION000025
种质名称：无刺卡因（巴冲）
原产地：泰国
资源类型：引进品种
主要用途：鲜食或加工
种质来源地：2007年中国热带农业科学院南亚热带作物研究所从泰国引进
植株姿态：开张
定植期：2021年2月下旬
现红期：2022年4月中旬
营养生长期：约420 d
初花期：2022年4月下旬
花开放时间：约20 d
成熟期：2022年8月上旬至9月下旬
果实发育期：约125 d
果实形状：圆筒形
单果质量：927 g
纵径：13.2 cm
横径：10.7 cm
果形指数：1.2

未成熟果实果皮颜色：暗绿色
成熟果实果皮颜色：暗黄/深黄色
果颈：无
果基：平
果顶：平
果实小果能否剥离：不可剥离
果眼外观：扁平或微凹
果眼深度：浅
果眼排列方式：右旋
果瘤：无
果肉颜色：淡黄色
果实香味：清香/微香
果实风味：酸甜
果肉质地：滑、脆/爽脆
果实外观综合评价：中
果实品质综合评价：中
果肉可溶性固形物含量：13.4%
果实成熟特性：晚熟
成熟期的一致性：不一致
丰产性：一般

Smooth Cayenne (Pakchong)

Code: ACCESSION000025
Name: Smooth Cayenne（Pakchong）
Place of origin: Thailand
Resource type: introduced variety
Main ways of consumption: fresh or processing
Initially introduction place: SSCRI, CATAS introduced from Thailand in 2007
Plant posture: spreading
Planting time: late February in 2021
Open heart stage: mid-April in 2022
Vegetative growth period: approx. 420 d
Initial flowering time: late April in 2022
Flowering period: approx. 20 d
Fruit maturation time: from early August to late September in 2022
Fruit developing period: approx. 125 d
Fruit shape: cylindrical
Single fruit weight: 927 g
Longitudinal diameter: 13.2 cm
Transverse diameter: 10.7 cm
Fruit shape index: 1.2
Immature fruit peel color: dark green
Mature fruit peel color: dark yellow/deep yellow
Fruit neck: absent
Fruit base: flat
Fruit top: flat
Fruitlets adhesion situation: inseparable
Fruit eyes appearance: flat or slightly concave
Fruit eyes depth: shallow
Fruit eyes arrangement: dextrorotation
Fruit tumors: absent
Flesh color: pale yellow
Fruit aroma: faint/slight scent
Fruit flavour: sour-sweet
Flesh texture: smooth, crisp/crunchy
Fruit appearance comprehensive evaluation: medium
Fruit quality comprehensive evaluation: medium
Soluble solids content in flesh: 19.2 %
Fruit ripening characteristics: late mature
Maturity consistency: nonuniform
Productivity: normal level

菠萝种质资源图谱 （上册）
Pineapple Germplasm Resources Map　(Volume 1)

植株
Plant

现红
Open heart

花
Flower

花序
Inflorescence

第三章 国外菠萝种质资源
Chapter 3 Pineapple Germplasm Resources from Abroad

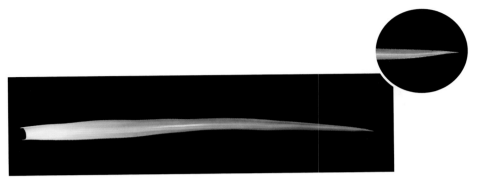

叶片
Leaf

带冠芽果
Fruit with crown bud

冠芽叶刺
Leaf spines in crown bud

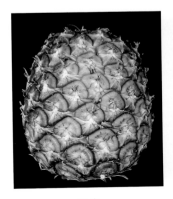

果实
Fruit

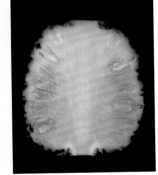

果实纵切
Fruit longitudinal section

果实横切
Fruit transverse section

泰国菠萝

编号：ACCESSION000071
种质名称：泰国菠萝
原产地：泰国
资源类型：引进品种
主要用途：鲜食或加工
种质来源地：2009年中国热带农业科学院南亚热带作物研究所从广东省内收集
植株姿态：开张
定植期：2021年2月下旬
现红期：2022年2月中旬
营养生长期：约360 d
初花期：2022年3月中旬
花开放时间：约20 d
成熟期：2022年6月中旬
果实发育期：约90 d
果实形状：（近）圆球形
单果质量：410 g
纵径：9.5 cm
横径：9.0 cm
果形指数：1.1

未成熟果实果皮颜色：银绿色
成熟果实果皮颜色：金黄/鲜黄色
果颈：无
果基：平
果顶：平
果实小果能否剥离：可剥离
果眼外观：微隆起
果眼深度：深
果眼排列方式：左旋或右旋
果瘤：无
果肉颜色：淡黄色
果实香味：清香/微香
果实风味：酸甜
果肉质地：脆/爽脆
果实外观综合评价：优
果实品质综合评价：中
果肉可溶性固形物含量：14.1%
果实成熟特性：中熟
成熟期的一致性：一致
丰产性：低产

Thailand Pineapple

Code: ACCESSION000071
Name: Thailand Pineapple
Place of origin: Thailand
Resource type: introduced variety
Main ways of consumption: fresh or processing
Initially introduction place: SSCRI, CATAS collected from Guangdong in 2009
Plant posture: spreading
Planting time: late February in 2021
Open heart stage: mid-February in 2022
Vegetative growth period: approx. 360 d
Initial flowering time: mid-March in 2022
Flowering period: approx. 20 d
Fruit maturation time: mid-June in 2022
Fruit developing period: approx. 90 d
Fruit shape: (near) spherical
Single fruit weight: 410 g
Longitudinal diameter: 9.5 cm
Transverse diameter: 9.0 cm
Fruit shape index: 1.1
Immature fruit peel color: silver green
Mature fruit peel color: golden yellow/vivid yellow
Fruit neck: absent
Fruit base: flat
Fruit top: flat
Fruitlets adhesion situation: separable
Fruit eyes appearance: slightly embossed
Fruit eyes depth: deep
Fruit eyes arrangement: levorotation or dextrorotation
Fruit tumors: absent
Flesh color: pale yellow
Fruit aroma: faint/slight scent
Fruit flavour: sour-sweet
Flesh texture: crisp/crunchy
Fruit appearance comprehensive evaluation: excellent
Fruit quality comprehensive evaluation: medium
Soluble solids content in flesh: 14.1 %
Fruit ripening characteristics: medium mature
Maturity consistency: uniform
Productivity: low

菠萝种质资源图谱 （上册）
Pineapple Germplasm Resources Map (Volume 1)

植株
Plant

现红
Open heart

花
Flower

花序
Inflorescence

第三章　国外菠萝种质资源
Chapter 3　Pineapple Germplasm Resources from Abroad

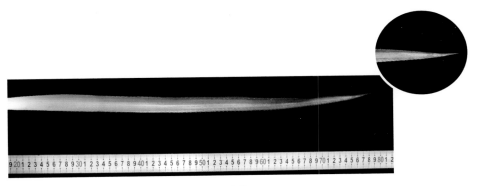

叶片
Leaf

带冠芽果
Fruit with crown bud

冠芽叶刺
Leaf spines in crown bud

果实
Fruit

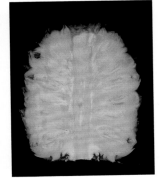

果实纵切
Fruit longitudinal section

果实横切
Fruit transverse section

Baro Rothchild

编号：ACCESSION000086
种质名称：Baro Rothchild
原产地：印度
资源类型：引进品种
主要用途：鲜食或加工
种质来源地：2008年中国热带农业科学院南亚热带作物研究所从印度引进
植株姿态：开张
定植期：2021年2月上旬
现红期：2022年3月上旬
营养生长期：约390 d
初花期：2022年3月下旬
花开放时间：约12 d
成熟期：2022年7月中旬至8月上旬
果实发育期：约120 d
果实形状：圆筒形
单果质量：1024 g
纵径：14.0 cm
横径：10.7 cm
果形指数：1.3

未成熟果实果皮颜色：暗绿色
成熟果实果皮颜色：黄色带绿斑
果颈：无
果基：平
果顶：平
果实小果能否剥离：不可剥离
果眼外观：扁平或微凹
果眼深度：浅
果眼排列方式：左旋或右旋
果瘤：无或多
果肉颜色：淡黄色
果实香味：无
果实风味：酸甜
果肉质地：脆/爽脆
果实外观综合评价：中
果实品质综合评价：好
果肉可溶性固形物含量：16.0%
果实成熟特性：晚熟
成熟期的一致性：不一致
丰产性：一般

Baro Rothchild

Code: ACCESSION000086
Name: Baro Rothchild
Place of origin: India
Resource type: introduced variety
Main ways of consumption: fresh or processing
Initially introduction place: SSCRI, CATAS introduced from India in 2008
Plant posture: spreading
Planting time: early February in 2021
Open heart stage: early March in 2022
Vegetative growth period: approx. 390 d
Initial flowering time: late March in 2022
Flowering period: approx. 12 d
Fruit maturation time: from mid-July to early August in 2022
Fruit developing period: approx. 120 d
Fruit shape: cylindrical
Single fruit weight: 1024 g
Longitudinal diameter: 14.0 cm
Transverse diameter: 10.7 cm
Fruit shape index: 1.3
Immature fruit peel color: dark green
Mature fruit peel color: yellow with green stripe
Fruit neck: absent
Fruit base: flat
Fruit top: flat
Fruitlets adhesion situation: inseparable
Fruit eyes appearance: flat or slightly concave
Fruit eyes depth: shallow
Fruit eyes arrangement: levorotation or dextrorotation
Fruit tumors: absent or many
Flesh color: pale yellow
Fruit aroma: no scent
Fruit flavour: sour-sweet
Flesh texture: crisp/crunchy
Fruit appearance comprehensive evaluation: medium
Fruit quality comprehensive evaluation: good
Soluble solids content in flesh: 16.0 %
Fruit ripening characteristics: late mature
Maturity consistency: nonuniform
Productivity: normal level

菠萝种质资源图谱 （上册）
Pineapple Germplasm Resources Map (Volume 1)

植株
Plant

现红
Open heart

花
Flower

花序
Inflorescence

第三章 国外菠萝种质资源
Chapter 3　Pineapple Germplasm Resources from Abroad

叶片
Leaf

带冠芽果
Fruit with crown bud

冠芽叶刺
Leaf spines in crown bud

果实
Fruit

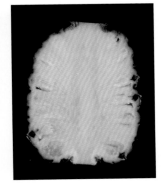

果实纵切
Fruit longitudinal section

果实横切
Fruit transverse section

C180

编号：ACCESSION000091
种质名称：C180
原产地：印度
资源类型：引进品种
主要用途：鲜食或加工
种质来源地：2008 年中国热带农业科学院南亚热带作物研究所从印度引进
植株姿态：开张
定植期：2021 年 2 月上旬
现红期：2022 年 2 月下旬
营养生长期：约 365 d
初花期：2022 年 3 月下旬
花开放时间：约 20 d
成熟期：2022 年 8 月上旬至 8 月下旬
果实发育期：约 140 d
果实形状：圆筒形
单果质量：1011 g
纵径：14.0 cm
横径：10.9 cm
果形指数：1.3

未成熟果实果皮颜色：淡绿色
成熟果果皮颜色：亮黄色
果颈：无
果基：突起
果顶：平
果实小果能否剥离：可剥离
果眼外观：扁平或微凹
果眼深度：较浅
果眼排列方式：右旋
果瘤：多
果肉颜色：淡黄色
果实香味：无
果实风味：甜酸
果肉质地：脆 / 爽脆
果实外观综合评价：中
果实品质综合评价：中
果肉可溶性固形物含量：13.3%
果实成熟特性：晚熟
成熟期的一致性：不一致
丰产性：一般

C180

Code: ACCESSION000091
Name: C180
Place of origin: India
Resource type: introduced variety
Main ways of consumption: fresh or processing
Initially introduction place: SSCRI, CATAS introduced from India in 2008
Plant posture: spreading
Planting time: early February in 2021
Open heart stage: late February in 2022
Vegetative growth period: approx. 365 d
Initial flowering time: late March in 2022
Flowering period: approx. 20 d
Fruit maturation time: early-to-late August in 2022
Fruit developing period: approx. 140 d
Fruit shape: cylindrical
Single fruit weight: 1011 g
Longitudinal diameter: 14.0 cm
Transverse diameter: 10.9 cm
Fruit shape index: 1.3
Immature fruit peel color: light green
Mature fruit peel color: bright yellow
Fruit neck: absent
Fruit base: protrusive
Fruit top: flat
Fruitlets adhesion situation: separable
Fruit eyes appearance: flat or slightly concave
Fruit eyes depth: relatively shallow
Fruit eyes arrangement: dextrorotation
Fruit tumors: many
Flesh color: pale yellow
Fruit aroma: no scent
Fruit flavour: sweet-sour
Flesh texture: crisp/crunchy
Fruit appearance comprehensive evaluation: medium
Fruit quality comprehensive evaluation: medium
Soluble solids content in flesh: 13.3 %
Fruit ripening characteristics: late mature
Maturity consistency: nonuniform
Productivity: normal level

菠萝种质资源图谱 （上册）
Pineapple Germplasm Resources Map (Volume 1)

植株
Plant

现红
Open heart

花
Flower

花序
Inflorescence

第三章　国外菠萝种质资源
Chapter 3　Pineapple Germplasm Resources from Abroad

叶片
Leaf

带冠芽果
Fruit with crown bud

冠芽叶刺
Leaf spines in crown bud

果实
Fruit

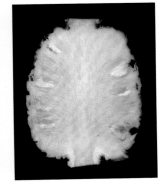

果实纵切
Fruit longitudinal section

果实横切
Fruit transverse section

Charlotte Rothschild

编号：ACCESSION000090
种质名称：Charlotte Rothschild
原产地：印度
资源类型：引进品种
主要用途：鲜食或加工
种质来源地：2008年中国热带农业科学院南亚热带作物研究所从印度引进
植株姿态：开张
定植期：2021年3月上旬
现红期：2022年3月下旬
营养生长期：约380 d
初花期：2022年4月下旬
花开放时间：约20 d
成熟期：2022年7月中旬至9月下旬
果实发育期：约125 d
果实形状：圆筒形
单果质量：1008 g
纵径：14.5 cm
横径：11.1 cm
果形指数：1.3

未成熟果实果皮颜色：绿色
成熟果实果皮颜色：浅褐色
果颈：有
果基：平
果顶：平
果实小果能否剥离：不可剥离
果眼外观：扁平或微凹
果眼深度：浅
果眼排列方式：左旋
果瘤：无
果肉颜色：淡黄色
果实香味：无
果实风味：酸甜
果肉质地：粗糙
果实外观综合评价：中
果实品质综合评价：差
果肉可溶性固形物含量：12.7%
果实成熟特性：中晚熟
成熟期的一致性：不一致
丰产性：一般

Charlotte Rothschild

Code: ACCESSION000090
Name: Charlotte Rothschild
Place of origin: India
Resource type: introduced variety
Main ways of consumption: fresh or processing
Initially introduction place: SSCRI, CATAS introduced from India in 2008
Plant posture: spreading
Planting time: early March in 2021
Open heart stage: late March in 2022
Vegetative growth period: approx. 380 d
Initial flowering time: late April in 2022
Flowering period: approx. 20 d
Fruit maturation time: from mid-July to late September in 2022
Fruit developing period: approx. 125 d
Fruit shape: cylindrical
Single fruit weight: 1008 g
Longitudinal diameter: 14.5 cm
Transverse diameter: 11.1 cm
Fruit shape index: 1.3
Immature fruit peel color: green
Mature fruit peel color: light brown
Fruit neck: present
Fruit base: flat
Fruit top: flat
Fruitlets adhesion situation: inseparable
Fruit eyes appearance: flat or slightly concave
Fruit eyes depth: shallow
Fruit eyes arrangement: levorotation
Fruit tumors: absent
Flesh color: pale yellow
Fruit aroma: no scent
Fruit flavour: sour-sweet
Flesh texture: coarse
Fruit appearance comprehensive evaluation: medium
Fruit quality comprehensive evaluation: poor
Soluble solids content in flesh: 12.7 %
Fruit ripening characteristics: medium-late mature
Maturity consistency: nonuniform
Productivity: normal level

菠萝种质资源图谱 （上册）
Pineapple Germplasm Resources Map　(Volume 1)

植株
Plant

现红
Open heart

花
Flower

花序
Inflorescence

第三章 国外菠萝种质资源
Chapter 3 Pineapple Germplasm Resources from Abroad

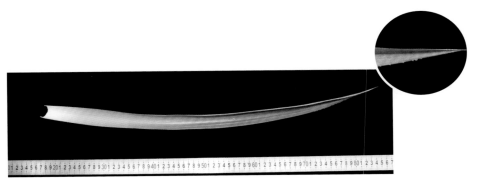

叶片
Leaf

带冠芽果
Fruit with crown bud

冠芽叶刺
Leaf spines in crown bud

果实
Fruit

果实纵切
Fruit longitudinal section

果实横切
Fruit transverse section

Kallara 土种

编号：ACCESSION000017
种质名称：Kallara 土种
原产地：印度
资源类型：引进品种
主要用途：鲜食或加工
种质来源地：2007 年中国热带农业科学院南亚热带作物研究所从印度引进
植株姿态：开张
定植期：2021 年 2 月中旬
现红期：2022 年 2 月中旬
营养生长期：约 360 d
初花期：2022 年 3 月中下旬
花开放时间：约 21 d
成熟期：2022 年 6 月上中旬
果实发育期：约 80 d
果实形状：圆筒形
单果质量：684 g
纵径：13.1 cm
横径：9.6 cm
果形指数：1.4

未成熟果实果皮颜色：暗绿色
成熟果实果皮颜色：金黄色
果颈：无
果基：弧形
果顶：平
果实小果能否剥离：可剥离
果眼外观：突起 / 隆起
果眼深度：深
果眼排列方式：左旋或右旋
果瘤：无
果肉颜色：淡黄
果实香味：清香 / 微香
果实风味：清甜
果肉质地：脆 / 爽脆
果实外观综合评价：好
果实品质综合评价：好
果肉可溶性固形物含量：16.0%
果实成熟特性：中熟
成熟期的一致性：基本一致
丰产性：低产

Kallara Local

Code: ACCESSION000017
Name: Kallara Local
Place of origin: India
Resource type: introduced variety
Main ways of consumption: fresh or processing
Initially introduction place: SSCRI, CATAS introduced from India in 2007
Plant posture: spreading
Planting time: mid-February in 2021
Open heart stage: mid-February in 2022
Vegetative growth period: approx. 360 d
Initial flowering time: mid-to-late March in 2022
Flowering period: approx. 21 d
Fruit maturation time: early-to-mid June in 2022
Fruit developing period: approx. 80 d
Fruit shape: cylindrical
Single fruit weight: 684 g
Longitudinal diameter: 13.1 cm
Transverse diameter: 9.6 cm
Fruit shape index: 1.4
Immature fruit peel color: dark green
Mature fruit peel color: golden yellow
Fruit neck: absent
Fruit base: curved
Fruit top: flat
Fruitlets adhesion situation: separable
Fruit eyes appearance: convex/embossed
Fruit eyes depth: deep
Fruit eyes arrangement: levorotation or dextrorotation
Fruit tumors: absent
Flesh color: pale yellow
Fruit aroma: faint/slight scent
Fruit flavour: mildly sweet
Flesh texture: crisp/crunchy
Fruit appearance comprehensive evaluation: good
Fruit quality comprehensive evaluation: good
Soluble solids content in flesh: 16.0 %
Fruit ripening characteristics: medium mature
Maturity consistency: basically uniform
Productivity: poor

植株
Plant

现红
Open heart

花
Flower

花序
Inflorescence

第三章 国外菠萝种质资源
Chapter 3 Pineapple Germplasm Resources from Abroad

叶片
Leaf

带冠芽果
Fruit with crown bud

冠芽叶刺
Leaf spines in crown bud

果实
Fruit

果实纵切
Fruit longitudinal section

果实横切
Fruit transverse section

詹姆斯皇后

编号：ACCESSION000087
种质名称：詹姆斯皇后
原产地：印度
资源类型：引进品种
主要用途：鲜食或加工
种质来源地：2008 年中国热带农业科学院南亚热带作物研究所从印度引进
植株姿态：开张
定植期：2021 年 2 月上旬
现红期：2022 年 2 月上旬
营养生长期：约 365 d
初花期：2022 年 3 月中旬
花开放时间：约 19 d
成熟期：2022 年 5 月下旬至 6 月上旬
果实发育期：约 75 d
果实形状：圆筒形
单果质量：568 g
纵径：12.0 cm
横径：9.0 cm
果形指数：1.3

未成熟果实果皮颜色：淡绿 / 绿色
成熟果实果皮颜色：暗黄 / 深黄色
果颈：无
果基：突起
果顶：平
果实小果能否剥离：不可剥离
果眼外观：突起 / 隆起
果眼深度：深
果眼排列方式：左旋或右旋
果瘤：少
果肉颜色：淡黄色
果实香味：清香 / 微香
果实风味：酸甜
果肉质地：粗糙
果实外观综合评价：中
果实品质综合评价：中
果肉可溶性固形物含量：16.3%
果实成熟特性：中熟
成熟期的一致性：基本一致
丰产性：低产

James Queen

Code: ACCESSION000087
Name: James Queen
Place of origin: India
Resource type: introduced variety
Main ways of consumption: fresh or processing
Initially introduction place: SSCRI, CATAS introduced from India in 2008
Plant posture: spreading
Planting time: early February in 2021
Open heart stage: early February in 2022
Vegetative growth period: approx. 365 d
Initial flowering time: mid-March in 2022
Flowering period: approx. 19 d
Fruit maturation time: from late May to early June in 2022
Fruit developing period: approx. 75 d
Fruit shape: cylindrical
Single fruit weight: 568 g
Longitudinal diameter: 12.0 cm
Transverse diameter: 9.0 cm
Fruit shape index: 1.3
Immature fruit peel color: light green/green
Mature fruit peel color: dark yellow/deep yellow
Fruit neck: absent
Fruit base: protrusive
Fruit top: flat
Fruitlets adhesion situation: inseparable
Fruit eyes appearance: convex/embossed
Fruit eyes depth: deep
Fruit eyes arrangement: levorotation or dextrorotation
Fruit tumors: few
Flesh color: pale yellow
Fruit aroma: faint/slight scent
Fruit flavour: sour-sweet
Flesh texture: coarse
Fruit appearance comprehensive evaluation: medium
Fruit quality comprehensive evaluation: medium
Soluble solids content in flesh: 16.3 %
Fruit ripening characteristics: medium mature
Maturity consistency: basically uniform
Productivity: low

菠萝种质资源图谱 （上册）
Pineapple Germplasm Resources Map (Volume 1)

植株
Plant

现红
Open heart

花
Flower

花序
Inflorescence

第三章 国外菠萝种质资源
Chapter 3 Pineapple Germplasm Resources from Abroad

叶片
Leaf

带冠芽果
Fruit with crown bud

冠芽叶刺
Leaf spines in crown bud

果实
Fruit

果实纵切
Fruit longitudinal section

果实横切
Fruit transverse section

印度皇后

编号：ACCESSION000088
种质名称：印度皇后
原产地：印度
资源类型：引进品种
主要用途：鲜食或加工
种质来源地：2008年中国热带农业科学院南亚热带作物研究所从印度引进
植株姿态：开张
定植期：2021年2月中旬
现红期：2022年2月中旬
营养生长期：约360 d
初花期：2022年3月中旬
花开放时间：约21 d
成熟期：2022年6月上中旬
果实发育期：约85 d
果实形状：圆筒形
单果质量：484 g
纵径：11.1 cm
横径：9.1 cm
果形指数：1.2

未成熟果实果皮颜色：暗绿色
成熟果实果皮颜色：暗黄/深黄色
果颈：无
果基：弧形
果顶：平
果实小果能否剥离：可剥离
果眼外观：突起/隆起
果眼深度：深
果眼排列方式：左旋或右旋
果瘤：少
果肉颜色：黄色
果实香味：无
果实风味：清甜
果肉质地：脆/爽脆
果实外观综合评价：中
果实品质综合评价：好
果肉可溶性固形物含量：15.2%
果实成熟特性：中熟
成熟期的一致性：基本一致
丰产性：低产

Indian Queen

Code: ACCESSION000088
Name: Indian Queen
Place of origin: India
Resource type: introduced variety
Main ways of consumption: fresh or processing
Initially introduction place: SSCRI, CATAS introduced from India in 2008
Plant posture: spreading
Planting time: mid-February in 2021
Open heart stage: mid-February in 2022
Vegetative growth period: approx. 360 d
Initial flowering time: mid-March in 2022
Flowering period: approx. 21 d
Fruit maturation time: early-to-mid June in 2022
Fruit developing period: approx. 85 d
Fruit shape: cylindrical
Single fruit weight: 484 g
Longitudinal diameter: 11.1 cm
Transverse diameter: 9.1 cm
Fruit shape index: 1.2
Immature fruit peel color: dark green
Mature fruit peel color: dark yellow/deep yellow
Fruit neck: absent
Fruit base: curved
Fruit top: flat
Fruitlets adhesion situation: separable
Fruit eyes appearance: convex/ embossed
Fruit eyes depth: deep
Fruit eyes arrangement: levorotation or dextrorotation
Fruit tumors: few
Flesh color: yellow
Fruit aroma: no scent
Fruit flavour: mildly sweet
Flesh texture: crisp/crunchy
Fruit appearance comprehensive evaluation: medium
Fruit quality comprehensive evaluation: good
Soluble solids content in flesh: 15.2 %
Fruit ripening characteristics: medium mature
Maturity consistency: basically uniform
Productivity: low

菠萝种质资源图谱 （上册）
Pineapple Germplasm Resources Map (Volume 1)

植株
Plant

现红
Open heart

花
Flower

花序
Inflorescence

第三章 国外菠萝种质资源
Chapter 3 Pineapple Germplasm Resources from Abroad

叶片
Leaf

带冠芽果
Fruit with crown bud

冠芽叶刺
Leaf spines in crown bud

果实
Fruit

果实纵切
Fruit longitudinal section

果实横切
Fruit transverse section

印度引未知名 1

编号：ACCESSION000089
种质名称：印度引未知名 1
原产地：印度
资源类型：引进品种
主要用途：鲜食或加工
种质来源地：2008 年中国热带农业科学院南亚热带作物研究所从印度引进
植株姿态：开张
定植期：2021 年 2 月中旬
现红期：2022 年 2 月中旬
营养生长期：约 360 d
初花期：2022 年 3 月中旬
花开放时间：约 20 d
成熟期：2022 年 6 月上中旬
果实发育期：约 85 d
果实形状：长圆筒形
单果质量：638 g
纵径：12.9 cm
横径：9.4 cm
果形指数：1.4

未成熟果实果皮颜色：暗绿色
成熟果实果皮颜色：金黄/鲜黄色
果颈：无
果基：突起
果顶：平
果实小果能否剥离：可剥离
果眼外观：突起/隆起
果眼深度：深
果眼排列方式：左旋或右旋
果瘤：无
果肉颜色：淡黄色
果实香味：清香/微香
果实风味：甜酸
果肉质地：脆/爽脆
果实外观综合评价：优
果实品质综合评价：好
果肉可溶性固形物含量：14.2%
果实成熟特性：中熟
成熟期的一致性：基本一致
丰产性：低产

Unknown cultivar No. 1 from Indian

Code: ACCESSION000089
Name: Unknown cultivar No. 1 from Indian
Place of origin: India
Resource type: introduced variety
Main ways of consumption: fresh or processing
Initially introduction place: SSCRI, CATAS introduced from India in 2008
Plant posture: spreading
Planting time: mid-February in 2021
Open heart stage: mid-February in 2022
Vegetative growth period: approx. 360 d
Initial flowering time: mid-March in 2022
Flowering period: approx. 20 d
Fruit maturation time: early-to-mid June in 2022
Fruit developing period: approx. 85 d
Fruit shape: long cylindrical
Single fruit weight: 638 g
Longitudinal diameter: 12.9 cm
Transverse diameter: 9.4 cm
Fruit shape index: 1.4
Immature fruit peel color: dark green
Mature fruit peel color: golden yellow/vivid yellow
Fruit neck: absent
Fruit base: protrusive
Fruit top: flat
Fruitlets adhesion situation: separable
Fruit eyes appearance: convex/embossed
Fruit eyes depth: deep
Fruit eyes arrangement: levorotation or dextrorotation
Fruit tumors: absent
Flesh color: pale yellow
Fruit aroma: faint/slight scent
Fruit flavour: sweet-sour
Flesh texture: crisp/crunchy
Fruit appearance comprehensive evaluation: excellent
Fruit quality comprehensive evaluation: good
Soluble solids content in flesh: 14.2 %
Fruit ripening characteristics: medium mature
Maturity consistency: basically uniform
Productivity: low

菠萝种质资源图谱 （上册）
Pineapple Germplasm Resources Map　(Volume 1)

植株
Plant

现红
Open heart

花
Flower

花序
Inflorescence

第三章　国外菠萝种质资源
Chapter 3 Pineapple Germplasm Resources from Abroad

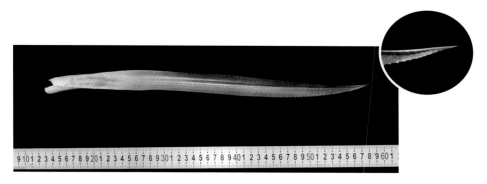

叶片
Leaf

带冠芽果
Fruit with crown bud

冠芽叶刺
Leaf spines in crown bud

果实
Fruit

果实纵切
Fruit longitudinal section

果实横切
Fruit transverse section

印度引未知名 2

编号：ACCESSION000092
种质名称：印度引未知名 2
原产地：印度
资源类型：引进品种
主要用途：鲜食或加工
种质来源地：2008 年中国热带农业科学院南亚热带作物研究所从印度引进
植株姿态：开张
定植期：2021 年 2 月中旬
现红期：2022 年 2 月中旬
营养生长期：约 360 d
初花期：2022 年 3 月中下旬
花开放时间：约 20 d
成熟期：2022 年 6 月上旬至 6 月下旬
果实发育期：约 85 d
果实形状：圆筒形
单果质量：416 g
纵径：10.1 cm
横径：8.9 cm
果形指数：1.1

未成熟果实果皮颜色：暗墨绿色
成熟果实果皮颜色：金黄色
果颈：无
果基：平
果顶：平
果实小果能否剥离：可剥离
果眼外观：突起/隆起
果眼深度：较深
果眼排列方式：左旋或右旋
果瘤：无
果肉颜色：淡黄色
果实香味：清香/微香
果实风味：甜酸
果肉质地：脆/爽脆
果实外观综合评价：中
果实品质综合评价：好
果肉可溶性固形物含量：15.4%
果实成熟特性：中熟
成熟期的一致性：不一致
丰产性：低产

Unknown cultivar No. 2 from Indian

Code: ACCESSION000092
Name: Unknown cultivar No. 2 from Indian
Place of origin: India
Resource type: introduced variety
Main ways of consumption: fresh or processing
Initially introduction place: SSCRI, CATAS introduced from India in 2008
Plant posture: spreading
Planting time: mid-February in 2021
Open heart stage: mid-February in 2022
Vegetative growth period: approx. 360 d
Initial flowering time: mid-to-late March in 2022
Flowering period: approx. 20 d
Fruit maturation time: early-to-late June in 2022
Fruit developing period: approx. 85 d
Fruit shape: cylindrical
Single fruit weight: 416 g
Longitudinal diameter: 10.1 cm
Transverse diameter: 8.9 cm
Fruit shape index: 1.1
Immature fruit peel color: dark blackish green
Mature fruit peel color: golden yellow
Fruit neck: absent
Fruit base: flat
Fruit top: flat
Fruitlets adhesion situation: separable
Fruit eyes appearance: convex/embossed
Fruit eyes depth: relatively deep
Fruit eyes arrangement: levorotation or dextrorotation
Fruit tumors: absent
Flesh color: pale yellow
Fruit aroma: faint/slight scent
Fruit flavour: sweet-sour
Flesh texture: crisp/crunchy
Fruit appearance comprehensive evaluation: medium
Fruit quality comprehensive evaluation: good
Soluble solids content in flesh: 15.4 %
Fruit ripening characteristics: medium mature
Maturity consistency: nonuniform
Productivity: low

菠萝种质资源图谱 （上册）
Pineapple Germplasm Resources Map (Volume 1)

植株
Plant

现红
Open heart

花
Flower

花序
Inflorescence

第三章　国外菠萝种质资源
Chapter 3　Pineapple Germplasm Resources from Abroad

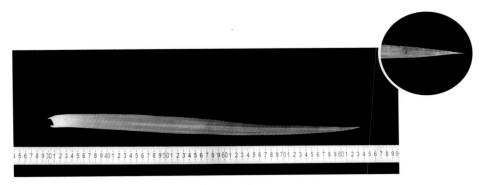

叶片
Leaf

带冠芽果
Fruit with crown bud

冠芽叶刺
Leaf spines in crown bud

果实
Fruit

果实纵切
Fruit longitudinal section

果实横切
Fruit transverse section

N36

编号：ACCESSION000049
种质名称：N36
原产地：马来西亚
资源类型：引进品种
主要用途：鲜食或加工
种质来源地：2008年中国热带农业科学院南亚热带作物研究所从马来西亚引进
系谱：Gandul（Singapore Spanish）× Sarawak 的杂交种
选育单位：马来西亚农业研究发展研究院
植株姿态：开张
定植期：2021 年 2 月上旬
现红期：2022 年 3 月上旬
营养生长期：约 390 d
初花期：2022 年 3 月下旬
花开放时间：约 31 d
成熟期：2022 年 7 月上旬至 7 月下旬
果实发育期：约 110 d
果实形状：长圆筒形
单果质量：1406 g
纵径：16.2 cm

横径：11.6 cm
果形指数：1.4
未成熟果实果皮颜色：淡绿 / 绿色
成熟果实果皮颜色：金黄 / 鲜黄色
果颈：无
果基：弧形
果顶：平
果实小果能否剥离：不可剥离
果眼外观：扁平或微凹
果眼深度：深
果眼排列方式：左旋
果瘤：多
果肉颜色：淡黄色
果实香味：无
果实风味：酸甜
果肉质地：粗糙
果实外观综合评价：中
果实品质综合评价：中
果肉可溶性固形物含量：14.3%
果实成熟特性：晚熟
成熟期的一致性：不一致
丰产性：丰产

N36

Code: ACCESSION000049

Name: N36

Place of origin: Malaysia

Resource type: introduced variety

Main ways of consumption: fresh or processing

Initially introduction place: SSCRI, CATAS introduced from Malaysia in 2008

Lineage: a cross between Gandul (Singapore Spanish) and Sarawak

Breeding organization: Malaysian Agricultural Research and Development Institute

Plant posture: spreading

Planting time: early February in 2021

Open heart stage: early March in 2022

Vegetative growth period: approx. 390 d

Initial flowering time: late March in 2022

Flowering period: approx. 31 d

Fruit maturation time: early-to-late July in 2022

Fruit developing period: approx. 110 d

Fruit shape: long cylindrical

Single fruit weight: 1406 g

Longitudinal diameter: 16.2 cm

Transverse diameter: 11.6 cm

Fruit shape index: 1.4

Immature fruit peel color: light green/green

Mature fruit peel color: golden yellow/vivid yellow

Fruit neck: absent

Fruit base: curved

Fruit top: flat

Fruitlets adhesion situation: inseparable

Fruit eyes appearance: flat or slightly concave

Fruit eyes depth: deep

Fruit eyes arrangement: levorotation

Fruit tumors: many

Flesh color: pale yellow

Fruit aroma: no scent

Fruit flavour: sour-sweet

Flesh texture: coarse

Fruit appearance comprehensive evaluation: medium

Fruit quality comprehensive evaluation: medium

Soluble solids content in flesh: 14.3 %

Fruit ripening characteristics: late mature

Maturity consistency: nonuniform

Productivity: productive

菠萝种质资源图谱 （上册）
Pineapple Germplasm Resources Map (Volume 1)

植株
Plant

现红
Open heart

花
Flower

花序
Inflorescence

第三章 国外菠萝种质资源
Chapter 3 Pineapple Germplasm Resources from Abroad

叶片
Leaf

带冠芽果
Fruit with crown bud

冠芽叶刺
Leaf spines in crown bud

果实
Fruit

果实纵切
Fruit longitudinal section

果实横切
Fruit transverse section

沙捞越

编号：ACCESSION000052
种质名称：沙捞越
原产地：马来西亚
资源类型：引进品种
主要用途：鲜食或加工
种质来源地：2008年中国热带农业科学院南亚热带作物研究所从马来西亚引进
植株姿态：开张
定植期：2021年2月下旬
现红期：2022年3月上旬
营养生长期：约370 d
初花期：2022年3月下旬
花开放时间：约22 d
成熟期：2022年7月下旬至8月中旬
果实发育期：约130 d
果实形状：长圆筒形
单果质量：803 g
纵径：14.2 cm
横径：9.7 cm
果形指数：1.5

未成熟果实果皮颜色：银绿色
成熟果实果皮颜色：暗黄/深黄色
果颈：有
果基：突起
果顶：钝圆
果实小果能否剥离：不可剥离
果眼外观：扁平或微凹
果眼深度：浅
果眼排列方式：右旋或其他
果瘤：多
果肉颜色：淡黄色
果实香味：清香/微香
果实风味：清甜
果肉质地：滑
果实外观综合评价：差
果实品质综合评价：中
果肉可溶性固形物含量：12.7%
果实成熟特性：晚熟
成熟期的一致性：不一致
丰产性：低产

Sarawak

Code: ACCESSION000052
Name: Sarawak
Place of origin: Malaysia
Resource type: introduced variety
Main ways of consumption: fresh or processing
Initially introduction place: SSCRI, CATAS introduced from Malaysia in 2008
Plant posture: spreading
Planting time: late February in 2021
Open heart stage: early March in 2022
Vegetative growth period: approx. 370 d
Initial flowering time: late March in 2022
Flowering period: approx. 22 d
Fruit maturation time: from late July to mid-August in 2022
Fruit developing period: approx. 130 d
Fruit shape: long cylindrical
Single fruit weight: 803 g
Longitudinal diameter: 14.2 cm
Transverse diameter: 9.7 cm
Fruit shape index: 1.5
Immature fruit peel color: silver green
Mature fruit peel color: dark yellow/deep yellow
Fruit neck: present
Fruit base: protrusive
Fruit top: bluntly round
Fruitlets adhesion situation: inseparable
Fruit eyes appearance: flat or slightly concave
Fruit eyes depth: shallow
Fruit eyes arrangement: dextrorotation or other
Fruit tumors: many
Flesh color: pale yellow
Fruit aroma: faint/slight scent
Fruit flavour: mildly sweet
Flesh texture: smooth
Fruit appearance comprehensive evaluation: poor
Fruit quality comprehensive evaluation: medium
Soluble solids content in flesh: 12.7 %
Fruit ripening characteristics: late mature
Maturity consistency: nonuniform
Productivity: low

菠萝种质资源图谱 （上册）
Pineapple Germplasm Resources Map （Volume 1）

植株
Plant

现红
Open heart

花
Flower

花序
Inflorescence

第三章 国外菠萝种质资源
Chapter 3 Pineapple Germplasm Resources from Abroad

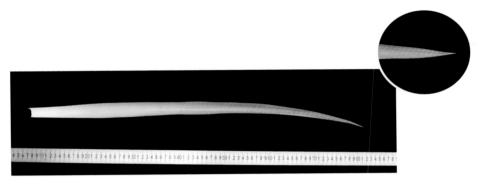

叶片
Leaf

带冠芽果
Fruit with crown bud

冠芽叶刺
Leaf spines in crown bud

果实
Fruit

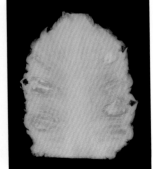

果实纵切
Fruit longitudinal section

果实横切
Fruit transverse section

Josapine

编号：ACCESSION000045
种质名称：Josapine，又名红香
原产地：马来西亚
资源类型：引进品种
主要用途：鲜食或加工
种质来源地：2008 年中国热带农业科学院南亚热带作物研究所从马来西亚引进
系谱：Johor（Singapore Spanish × Smooth Cayenne）× Sarawak 的杂交种
选育单位：马来西亚农业研究发展研究院
植株姿态：开张
定植期：2021 年 2 月下旬
现红期：2022 年 2 月中旬
营养生长期：约 355 d
初花期：2022 年 3 月中旬
花开放时间：约 25 d
成熟期：2022 年 5 月下旬
果实发育期：约 70 d
果实形状：圆筒形
单果质量：429 g
纵径：9.1 cm
横径：8.8 cm
果形指数：1.0
未成熟果实果皮颜色：暗紫红色
成熟果实果皮颜色：橘红 / 橙红色
果颈：无
果基：平
果顶：平
果实小果能否剥离：不可剥离
果眼外观：微隆起
果眼深度：较深
果眼排列方式：左旋、右旋或其他
果瘤：无
果肉颜色：黄色
果实香味：清香 / 微香
果实风味：甜酸
果肉质地：滑
果实外观综合评价：中
果实品质综合评价：好
果肉可溶性固形物含量：17.6%
果实成熟特性：早熟
成熟期的一致性：不一致
丰产性：低产

Chapter 3 Pineapple Germplasm Resources from Abroad

Josapine

Code: ACCESSION000045

Name: Josapine, also known as Hongxiang

Place of origin: Malaysia

Resource type: introduced variety

Main ways of consumption: fresh or processing

Initially introduction place: SSCRI, CATAS introduced from Malaysia in 2008

Lineage: a hybrid between Johor (Singapore Spanish × Smooth Cayenne) and Sarawak

Breeding organization: Malaysian Agricultural Research and Development Institute

Plant posture: spreading

Planting time: late February in 2021

Open heart stage: mid-February in 2022

Vegetative growth period: approx. 355 d

Initial flowering time: mid-March in 2022

Flowering period: approx. 25 d

Fruit maturation time: late May in 2022

Fruit developing period: approx. 70 d

Fruit shape: cylindrical

Single fruit weight: 429 g

Longitudinal diameter: 9.1 cm

Transverse diameter: 8.8 cm

Fruit shape index: 1.0

Immature fruit peel color: violet red

Mature fruit peel color: orange red

Fruit neck: absent

Fruit base: flat

Fruit top: flat

Fruitlets adhesion situation: inseparable

Fruit eyes appearance: slightly embossed

Fruit eyes depth: relatively deep

Fruit eyes arrangement: levorotation, dextrorotation or others

Fruit tumors: absent

Flesh color: yellow

Fruit aroma: faint/slight scent

Fruit flavour: sweet-sour

Flesh texture: smooth

Fruit appearance comprehensive evaluation: medium

Fruit quality comprehensive evaluation: good

Soluble solids content in flesh: 17.6%

Fruit ripening characteristics: early mature

Maturity consistency: nonuniform

Productivity: low

菠萝种质资源图谱 （上册）
Pineapple Germplasm Resources Map (Volume 1)

植株
Plant

现红
Open heart

花
Flower

花序
Inflorescence

第三章 国外菠萝种质资源
Chapter 3 Pineapple Germplasm Resources from Abroad

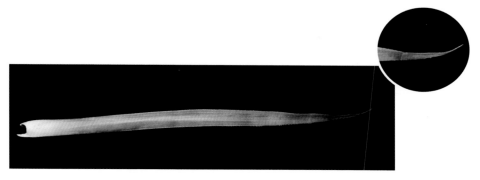

叶片
Leaf

带冠芽果
Fruit with crown bud

冠芽叶刺
Leaf spines in crown bud

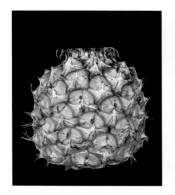

果实
Fruit

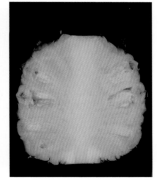

果实纵切
Fruit longitudinal section

果实横切
Fruit transverse section

观赏类绿果

编号：ACCESSION000101
种质名称：观赏类绿果
原产地：马来西亚
资源类型：引进品种
主要用途：鲜食或加工
种质来源地：2008年中国热带农业科学院南亚热带作物研究所从马来西亚引进
植株姿态：开张
定植期：2021年2月中旬
现红期：2022年2月下旬
营养生长期：约370 d
初花期：2022年4月上旬
花开放时间：约20 d
成熟期：2022年6月下旬至7月上旬
果实发育期：约85 d
果实形状：圆筒形
单果质量：590 g
纵径：11.1 cm
横径：9.4 cm
果形指数：1.2

未成熟果实果皮颜色：暗绿色
成熟果实果皮颜色：亮黄/淡黄色
果颈：无
果基：平
果顶：平
果实小果能否剥离：不可剥离
果眼外观：扁平或微凹
果眼深度：深
果眼排列方式：左旋或其他
果瘤：少
果肉颜色：淡黄色
果实香味：清香/微香
果实风味：酸
果肉质地：脆/爽脆
果实外观综合评价：中
果实品质综合评价：差
果肉可溶性固形物含量：11.3%
果实成熟特性：中熟
成熟期的一致性：基本一致
丰产性：低产

Ornamental Green Fruit

Code: ACCESSION000101
Name: Ornamental Green Fruit
Place of origin: Malaysia
Resource type: introduced variety
Main ways of consumption: fresh or processing
Place initially introduced: SSCRI, CATAS introduced from Malaysia in 2008
Plant posture: spreading
Planting time: mid-February in 2021
Open heart stage: late February in 2022
Vegetative growth period: approx. 370 d
Initial flowering time: early April in 2022
Flowering period: approx. 20 d
Fruit maturation time: from late June to early July in 2022
Fruit developing period: approx. 85 d
Fruit shape: cylindrical
Single fruit weight: 590 g
Longitudinal diameter: 11.1 cm
Transverse diameter: 9.4 cm
Fruit shape index: 1.2
Immature fruit peel color: dark green
Mature fruit peel color: bright yellow/light yellow
Fruit neck: absent
Fruit base: flat
Fruit top: flat
Fruitlets adhesion situation: inseparable
Fruit eyes appearance: flat or slightly concave
Fruit eyes depth: deep
Fruit eyes arrangement: levorotation or others
Fruit tumors: few
Flesh color: pale yellow
Fruit aroma: faint/slight scent
Fruit flavour: sour
Flesh texture: crisp/crunchy
Fruit appearance comprehensive evaluation: medium
Fruit quality comprehensive evaluation: poor
Soluble solids content in flesh: 11.3 %
Fruit ripening characteristics: medium mature
Maturity consistency: basically uniform
Productivity: low

菠萝种质资源图谱 （上册）
Pineapple Germplasm Resources Map　(Volume 1)

植株
Plant

现红
Open heart

花
Flower

花序
Inflorescence

第三章 国外菠萝种质资源
Chapter 3　Pineapple Germplasm Resources from Abroad

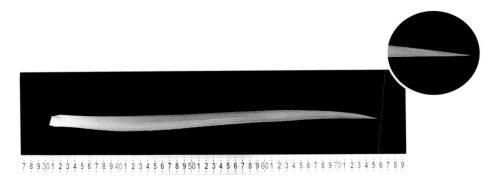

叶片
Leaf

带冠芽果
Fruit with crown bud

冠芽叶刺
Leaf spines in crown bud

果实
Fruit

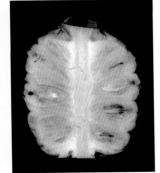

果实纵切
Fruit longitudinal section

果实横切
Fruit transverse section

马来西亚引未知名 1

编号：ACCESSION000125
种质名称：马来西亚引未知名 1
原产地：马来西亚
资源类型：引进品种
主要用途：鲜食或加工
种质来源地：2016 年中国热带农业科学院南亚热带作物研究所从海南引进
植株姿态：开张
定植期：2021 年 3 月上旬
现红期：2022 年 3 月下旬
营养生长期：约 380 d
初花期：2022 年 4 月上旬
花开放时间：约 20 d
成熟期：2022 年 7 月中旬至 8 月下旬
果实发育期：约 120 d
果实形状：圆筒形
单果质量：749 g
纵径：12.3 cm
横径：10.0 cm
果形指数：1.2

未成熟果实果皮颜色：淡绿/绿色
成熟果实果皮颜色：橘红/橙红色
果颈：无
果基：平
果顶：平
果实小果能否剥离：可剥离
果眼外观：微隆起
果眼深度：较深
果眼排列方式：左旋
果瘤：无
果肉颜色：奶油色
果实香味：清香/微香
果实风味：酸甜
果肉质地：粗糙
果实外观综合评价：中
果实品质综合评价：差
果肉可溶性固形物含量：11.1%
果实成熟特性：晚熟
成熟期的一致性：不一致
丰产性：低产

Unknown cultivar No. 1 from Malaysia

Code: ACCESSION000125
Name: Unknown cultivar No. 1 from Malaysia
Place of origin: Malaysia
Resource type: introduced variety
Main ways of consumption: fresh or processing
Initially introduction place: SSCRI, CATAS introduced from Malaysia in 2008
Plant posture: spreading
Planting time: early March in 2021
Open heart stage: late March in 2022
Vegetative growth period: approx. 380 d
Initial flowering time: early April in 2022
Flowering period: approx. 20 d
Fruit maturation time: from mid-July to late August in 2022
Fruit developing period: approx. 120 d
Fruit shape: cylindrical
Single fruit weight: 749 g
Longitudinal diameter: 12.3 cm
Vertical diameter: 10.0 cm
Fruit shape index: 1.2
Immature fruit peel color: light green/green
Mature fruit peel color: orange red
Fruit neck: absent
Fruit base: flat
Fruit top: flat
Fruitlets adhesion situation: separable
Fruit eyes appearance: slightly embossed
Fruit eyes depth: relatively deep
Fruit eyes arrangement: levorotation
Fruit tumors: absent
Flesh color: cream
Fruit aroma: faint/slight scent
Fruit flavour: sour-sweet
Flesh texture: coarse
Fruit appearance comprehensive evaluation: medium
Fruit quality comprehensive evaluation: poor
Soluble solids content in flesh: 11.1 %
Fruit ripening characteristics: late mature
Maturity consistency: nonuniform
Productivity: low

| 菠萝种质资源图谱 （上册）
Pineapple Germplasm Resources Map　(Volume 1)

植株
Plant

现红
Open heart

花
Flower

花序
Inflorescence

第三章 国外菠萝种质资源
Chapter 3 Pineapple Germplasm Resources from Abroad

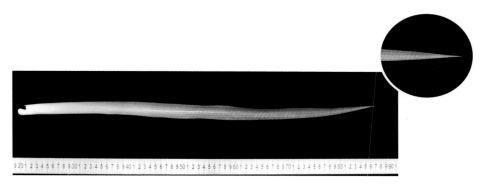

叶片
Leaf

带冠芽果
Fruit with crown bud

冠芽叶刺
Leaf spines in crown bud

果实
Fruit

果实纵切
Fruit longitudinal section

果实横切
Fruit transverse section

DL1

编号：ACCESSION000096
种质名称：DL1
原产地：越南
资源类型：引进品种
主要用途：鲜食或加工
种质来源地：2009年中国热带农业科学院南亚热带作物研究所从越南引进
植株姿态：开张
定植期：2021年2月下旬
现红期：2022年2月中旬
营养生长期：约355 d
初花期：2022年3月中旬
花开放时间：约21 d
成熟期：2022年6月中旬
果实发育期：约90 d
果实形状：长圆筒形
单果质量：585 g
纵径：13.3 cm
横径：8.8 cm
果形指数：1.5

未成熟果实果皮颜色：银绿色
成熟果实果皮颜色：亮黄/淡黄色
果颈：有
果基：弧形
果顶：平
果实小果能否剥离：可剥离
果眼外观：突起/隆起
果眼深度：深
果眼排列方式：左旋或右旋
果瘤：少或多
果肉颜色：淡黄色
果实香味：清香/微香
果实风味：酸甜
果肉质地：粗糙
果实外观综合评价：好
果实品质综合评价：好
果肉可溶性固形物含量：18.1%
果实成熟特性：中熟
成熟期的一致性：基本一致
丰产性：低产

DL1

Code: ACCESSION000096
Name: DL1
Place of origin: Vietnam
Resource type: introduced variety
Main ways of consumption: fresh or processing
Initially introduction place: SSCRI, CATAS introduced from Vietnam in 2009
Plant posture: spreading
Planting time: late February in 2021
Open heart stage: mid-February in 2022
Vegetative growth period: approx. 355 d
Initial flowering time: mid-March in 2022
Flowering period: approx. 21 d
Fruit maturation time: mid-June in 2022
Fruit developing period: approx. 90 d
Fruit shape: long cylindrical
Single fruit weight: 585 g
Longitudinal diameter: 13.3 cm
Transverse diameter: 8.8 cm
Fruit shape index: 1.5
Immature fruit peel color: silver green
Mature fruit peel color: bright yellow/light yellow
Fruit neck: present
Fruit base: curved
Fruit top: flat
Fruitlets adhesion situation: separable
Fruit eyes appearance: convex/embossed
Fruit eyes depth: deep
Fruit eyes arrangement: levorotation or dextrorotation
Fruit tumors: few or many
Flesh color: pale yellow
Fruit aroma: faint/slight scent
Fruit flavour: sour-sweet
Flesh texture: coarse
Fruit appearance comprehensive evaluation: good
Fruit quality comprehensive evaluation: good
Soluble solids content in flesh: 18.1 %
Fruit ripening characteristics: medium mature
Maturity consistency: basically uniform
Productivity: low

植株
Plant

现红
Open heart

花
Flower

花序
Inflorescence

第三章 国外菠萝种质资源
Chapter 3 Pineapple Germplasm Resources from Abroad

叶片
Leaf

带冠芽果
Fruit with crown bud

冠芽叶刺
Leaf spines in crown bud

果实
Fruit

果实纵切
Fruit longitudinal section

果实横切
Fruit transverse section

DL3

编号：ACCESSION000093
种质名称：DL3
原产地：越南
资源类型：引进品种
主要用途：鲜食或加工
种质来源地：2009年中国热带农业科学院南亚热带作物研究所从越南引进
植株姿态：开张
定植期：2021年2月下旬
现红期：2022年2月中旬
营养生长期：约355 d
初花期：2022年3月中旬
花开放时间：约22 d
成熟期：2022年6月上旬
果实发育期：约80 d
果实形状：长圆筒形
单果质量：619 g
纵径：13.8 cm
横径：9.1 cm
果形指数：1.5

未成熟果实果皮颜色：暗墨绿色
成熟果实果皮颜色：金黄色
果颈：有
果基：突起
果顶：钝圆
果实小果能否剥离：可剥离
果眼外观：突起/隆起
果眼深度：较深
果眼排列方式：左旋或右旋
果瘤：少
果肉颜色：黄色
果实香味：清香/微香
果实风味：清甜
果肉质地：脆/爽脆
果实外观综合评价：中
果实品质综合评价：中
果肉可溶性固形物含量：13.2%
果实成熟特性：中熟
成熟期的一致性：基本一致
丰产性：低产

DL3

Code: ACCESSION000093
Name: DL3
Place of origin: Vietnam
Resource type: introduced variety
Main ways of consumption: fresh or processing
Initially introduction place: SSCRI, CATAS introduced from Vietnam in 2009
Plant posture: spreading
Planting time: late February in 2021
Open heart stage: mid-February in 2022
Vegetative growth period: approx. 355 d
Initial flowering time: mid-March in 2022
Flowering period: approx. 22 d
Fruit maturation time: early June in 2022
Fruit developing period: approx. 80 d
Fruit shape: long cylindrical
Single fruit weight: 619 g
Longitudinal diameter: 13.8 cm
Transverse diameter: 9.1 cm
Fruit shape index: 1.5
Immature fruit peel color: dark blackish green
Mature fruit peel color: golden yellow
Fruit neck: present
Fruit base: protrusive
Fruit top: bluntly round
Fruitlets adhesion situation: separable
Fruit eyes appearance: convex/embossed
Fruit eyes depth: relatively deep
Fruit eyes arrangement: levorotation or dextrorotation
Fruit tumors: few
Flesh color: yellow
Fruit aroma: faint/slight scent
Fruit flavour: mildly sweet
Flesh texture: crisp/crunchy
Fruit appearance comprehensive evaluation: medium
Fruit quality comprehensive evaluation: medium
Soluble solids content in flesh: 13.2 %
Fruit ripening characteristics: medium mature
Maturity consistency: basically uniform
Productivity: low

菠萝种质资源图谱（上册）
Pineapple Germplasm Resources Map (Volume 1)

植株
Plant

现红
Open heart

花
Flower

花序
Inflorescence

第三章 国外菠萝种质资源
Chapter 3 Pineapple Germplasm Resources from Abroad

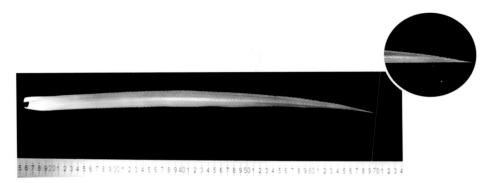

叶片
Leaf

带冠芽果
Fruit with crown bud

冠芽叶刺
Leaf spines in crown bud

果实
Fruit

果实纵切
Fruit longitudinal section

果实横切
Fruit transverse section

DN1

编号：ACCESSION000099
种质名称：DN1
原产地：越南
资源类型：引进品种
主要用途：鲜食或加工
种质来源地：2009年中国热带农业科学院南亚热带作物研究所从越南引进
植株姿态：开张
定植期：2021年2月上旬
现红期：2022年3月上旬
营养生长期：约390 d
初花期：2022年4月上旬
花开放时间：约21 d
成熟期：2022年7月上旬
果实发育期：约90 d
果实形状：圆筒形
单果质量：369 g
纵径：9.0 cm
横径：8.2 cm
果形指数：1.1

未成熟果实果皮颜色：暗墨绿色
成熟果实果皮颜色：亮黄/淡黄色
果颈：无
果基：平
果顶：平
果实小果能否剥离：不可剥离
果眼外观：扁平或微凹
果眼深度：较浅
果眼排列方式：其他
果瘤：无
果肉颜色：淡黄色
果实香味：清香/微香
果实风味：酸甜
果肉质地：脆/爽脆
果实外观综合评价：中
果实品质综合评价：差
果肉可溶性固形物含量：10.1%
果实成熟特性：中熟
成熟期的一致性：一致
丰产性：低产

DN1

Code: ACCESSION000099
Name: DN1
Place of origin: Vietnam
Resource type: introduced variety
Main ways of consumption: fresh or processing
Initially introduction place: SSCRI, CATAS introduced from Vietnam in 2009
Plant posture: spreading
Planting time: early February in 2021
Open heart stage: early March in 2022
Vegetative growth period: approx. 390 d
Initial flowering time: early April in 2022
Flowering period: approx. 21 d
Fruit maturation time: early July in 2022
Fruit developing period: approx. 90 d
Fruit shape: cylindrical
Single fruit weight: 369 g
Longitudinal diameter: 9.0 cm
Transverse diameter: 8.2 cm
Fruit shape index: 1.1
Immature fruit peel color: dark blackish green
Mature fruit peel color: bright yellow/light yellow
Fruit neck: absent
Fruit base: flat
Fruit top: flat
Fruitlets adhesion situation: inseparable
Fruit eyes appearance: flat or slightly concave
Fruit eyes depth: relatively shallow
Fruit eyes arrangement: other
Fruit tumors: absent
Flesh color: pale yellow
Fruit aroma: faint/slight scent
Fruit flavour: sour-sweet
Flesh texture: crisp/crunchy
Fruit appearance comprehensive evaluation: medium
Fruit quality comprehensive evaluation: poor
Soluble solids content in flesh: 10.1%
Fruit ripening characteristics: medium mature
Maturity consistency: uniform
Productivity: low

菠萝种质资源图谱 （上册）
Pineapple Germplasm Resources Map (Volume 1)

植株
Plant

现红
Open heart

花
Flower

花序
Inflorescence

第三章 国外菠萝种质资源
Chapter 3　Pineapple Germplasm Resources from Abroad

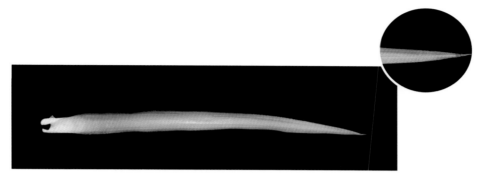

叶片
Leaf

带冠芽果
Fruit with crown bud

冠芽叶刺
Leaf spines in crown bud

果实
Fruit

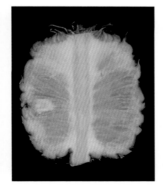

果实纵切
Fruit longitudinal section

果实横切
Fruit transverse section

DN2

编号：ACCESSION000102
种质名称：DN2
原产地：越南
资源类型：引进品种
主要用途：鲜食或加工
种质来源地：2009年中国热带农业科学院南亚热带作物研究所从越南引进
植株姿态：开张
定植期：2021年2月下旬
现红期：2022年2月下旬
营养生长期：约365 d
初花期：2022年3月中旬
花开放时间：约23 d
成熟期：2022年6月上中旬
果实发育期：约85 d
果实形状：长圆柱形
单果质量：502 g
纵径：12.2 cm
横径：8.7 cm
果形指数：1.4

未成熟果实果皮颜色：暗墨绿色
成熟果实果皮颜色：亮黄/淡黄色
果颈：无
果基：突起
果顶：平
果实小果能否剥离：可剥离
果眼外观：突起/隆起
果眼深度：较浅
果眼排列方式：左旋或右旋
果瘤：无
果肉颜色：淡黄色
果实香味：无
果实风味：甜酸
果肉质地：脆/爽脆
果实外观综合评价：中
果实品质综合评价：好
果肉可溶性固形物含量：16.1%
果实成熟特性：中熟
成熟期的一致性：基本一致
丰产性：低产

DN2

Code: ACCESSION000102
Name: DN2
Place of origin: Vietnam
Resource type: introduced variety
Main ways of consumption: fresh or processing
Initially introduction place: SSCRI, CATAS introduced from Vietnam in 2009
Plant posture: spreading
Planting time: late February in 2021
Open heart stage: late February in 2022
Vegetative growth period: approx. 365 d
Initial flowering time: mid-March in 2022
Flowering period: approx. 23 d
Fruit maturation time: early-to-mid June in 2022
Fruit developing period: approx. 85 d
Fruit shape: long cylindrical
Single fruit weight: 502 g
Longitudinal diameter: 12.2 cm
Transverse diameter: 8.7 cm
Fruit shape index: 1.4
Immature fruit peel color: dark blackish green
Mature fruit peel color: bright yellow/light yellow
Fruit neck: absent
Fruit base: protrusive
Fruit top: flat
Fruitlets adhesion situation: separable
Fruit eyes appearance: convex/embossed
Fruit eyes depth: relatively shallow
Fruit eyes arrangement: levorotation or dextrorotation
Fruit tumors: absent
Flesh color: pale yellow
Fruit aroma: no scent
Fruit flavour: sweet-sour
Flesh texture: crisp/crunchy
Fruit appearance comprehensive evaluation: medium
Fruit quality comprehensive evaluation: good
Soluble solids content in flesh: 16.1 %
Fruit ripening characteristics: medium mature
Maturity consistency: basically uniform
Productivity: low

菠萝种质资源图谱（上册）
Pineapple Germplasm Resources Map (Volume 1)

植株
Plant

现红
Open heart

花
Flower

花序
Inflorescence

第三章　国外菠萝种质资源
Chapter 3　Pineapple Germplasm Resources from Abroad

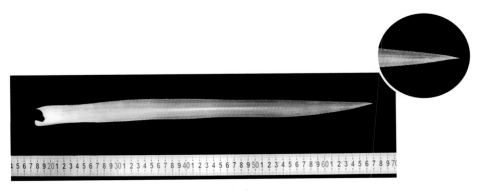

叶片
Leaf

带冠芽果
Fruit with crown bud

冠芽叶刺
Leaf spines in crown bud

果实
Fruit

果实纵切
Fruit longitudinal section

果实横切
Fruit transverse section

DN5

编号：ACCESSION000100
种质名称：DN5
原产地：越南
资源类型：引进品种
主要用途：鲜食或加工
种质来源地：2009 年中国热带农业科学院南亚热带作物研究所从越南引进
植株姿态：开张
定植期：2021 年 2 月上旬
现红期：2022 年 2 月上旬
营养生长期：约 365 d
初花期：2022 年 3 月中下旬
花开放时间：约 19 d
成熟期：2022 年 6 月上旬
果实发育期：约 75 d
果实形状：长圆筒形
单果质量：538 g
纵径：12.4 cm
横径：8.8 cm
果形指数：1.4

未成熟果实果皮颜色：暗绿
成熟果实果皮颜色：亮黄 / 淡黄色
果颈：无
果基：弧形
果顶：平
果实小果能否剥离：可剥离
果眼外观：突起 / 隆起
果眼深度：较浅
果眼排列方式：左旋或右旋
果瘤：无
果肉颜色：淡黄色
果实香味：清香 / 微香
果实风味：清甜
果肉质地：脆 / 爽脆
果实外观综合评价：中
果实品质综合评价：优
果肉可溶性固形物含量：17.7%
果实成熟特性：中熟
成熟期的一致性：一致
丰产性：低产

DN5

Code: ACCESSION000100
Name: DN5
Place of origin: Vietnam
Resource type: introduced variety
Main ways of consumption: fresh or processing
Initially introduction place: SSCRI, CATAS introduced from Vietnam in 2009
Plant posture: spreading
Planting time: early February in 2021
Open heart stage: early February in 2022
Vegetative growth period: approx. 365 d
Initial flowering time: mid-to-late March in 2022
Flowering period: approx. 19 d
Fruit maturation time: early June in 2022
Fruit developing period: approx. 75 d
Fruit shape: long cylindrical
Single fruit weight: 538 g
Longitudinal diameter: 12.4 cm
Transverse diameter: 8.8 cm
Fruit shape index: 1.4
Immature fruit peel color: dark green
Mature fruit peel color: bright yellow/light yellow
Fruit neck: absent
Fruit base: curved
Fruit top: flat
Fruitlets adhesion situation: separable
Fruit eyes appearance: convex/embossed
Fruit eyes depth: relatively shallow
Fruit eyes arrangement: levorotation or dextrorotation
Fruit tumors: absent
Flesh color: pale yellow
Fruit aroma: faint/slight scent
Fruit flavour: mildly sweet
Flesh texture: crisp/crunchy
Fruit appearance comprehensive evaluation: medium
Fruit quality comprehensive evaluation: excellent
Soluble solids content in flesh: 17.7 %
Fruit ripening characteristics: medium mature
Maturity consistency: uniform
Productivity: low

菠萝种质资源图谱 （上册）
Pineapple Germplasm Resources Map (Volume 1)

植株
Plant

现红
Open heart

花
Flower

花序
Inflorescence

第三章 国外菠萝种质资源
Chapter 3 Pineapple Germplasm Resources from Abroad

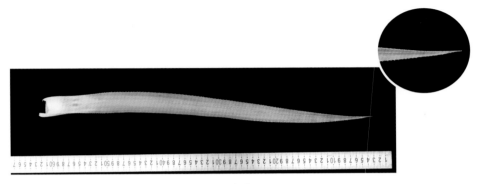

叶片
Leaf

带冠芽果
Fruit with crown bud

冠芽叶刺
Leaf spines in crown bud

果实
Fruit

果实纵切
Fruit longitudinal section

果实横切
Fruit transverse section

JPZ

编号：ACCESSION000095
种质名称：JPZ
原产地：越南
资源类型：引进品种
主要用途：鲜食或加工
种质来源地：2009年中国热带农业科学院南亚热带作物研究所从越南引进
植株姿态：开张
定植期：2021年2月下旬
现红期：2022年2月下旬
营养生长期：约365 d
初花期：2022年3月下旬
花开放时间：约22 d
成熟期：2022年6月下旬至7月中旬
果实发育期：约100 d
果实形状：圆筒形
单果质量：905 g
纵径：13.9 cm
横径：10.4 cm
果形指数：1.3

未成熟果实果皮颜色：淡绿色
成熟果实果皮颜色：金黄色
果颈：无
果基：平
果顶：平
果实小果能否剥离：不可剥离
果眼外观：扁平或微凹
果眼深度：较浅
果眼排列方式：左旋或右旋
果瘤：无或少
果肉颜色：黄色
果实香味：清香/微香
果实风味：酸甜
果肉质地：脆/爽脆
果实外观综合评价：中
果实品质综合评价：好
果肉可溶性固形物含量：18.2%
果实成熟特性：中晚熟
成熟期的一致性：基本一致
丰产性：一般

JPZ

Code: ACCESSION000095
Name: JPZ
Place of origin: Vietnam
Resource type: introduced variety
Main ways of consumption: fresh or processing
Initially introduction place: SSCRI, CATAS introduced from Vietnam in 2009
Plant posture: spreading
Planting time: late February in 2021
Open heart stage: late February in 2022
Vegetative growth period: approx. 365 d
Initial flowering time: late March in 2022
Flowering period: approx. 22 d
Fruit maturation time: from late June to mid-July in 2022
Fruit developing period: approx. 100 d
Fruit shape: cylindrical
Single fruit weight: 905 g
Longitudinal diameter: 13.9 cm
Vertical diameter: 10.4 cm
Fruit shape index: 1.3
Immature fruit peel color: light green
Mature fruit peel color: golden yellow
Fruit neck: absent
Fruit base: flat
Fruit top: flat
Fruitlets adhesion situation: inseparable
Fruit eyes appearance: flat or slightly concave
Fruit eyes depth: relatively shallow
Fruit eyes arrangement: levorotation or dextrorotation
Fruit tumors: absent or few
Flesh color: yellow
Fruit aroma: faint/slight scent
Fruit flavour: sour-sweet
Flesh texture: crisp/crunchy
Fruit appearance comprehensive evaluation: medium
Fruit quality comprehensive evaluation: good
Soluble solids content in flesh: 18.2 %
Fruit ripening characteristics: medium-late mature
Maturity consistency: basically uniform
Productivity: normal level

植株
Plant

现红
Open heart

花
Flower

花序
Inflorescence

第三章 国外菠萝种质资源
Chapter 3 Pineapple Germplasm Resources from Abroad

叶片
Leaf

带冠芽果
Fruit with crown bud

冠芽叶刺
Leaf spines in crown bud

果实
Fruit

果实纵切
Fruit longitudinal section

果实横切
Fruit transverse section

越南引皇后 1 号

编号：ACCESSION000043
种质名称：越南引皇后 1 号
原产地：越南
资源类型：引进品种
主要用途：鲜食或加工
种质来源地：2009 年中国热带农业科学院南亚热带作物研究所从越南引进
植株姿态：开张
定植期：2021 年 2 月下旬
现红期：2022 年 2 月中旬
营养生长期：约 360 d
初花期：2022 年 3 月中旬
花开放时间：约 24 d
成熟期：2022 年 6 月上旬
果实发育期：约 80 d
果实形状：长圆筒形
单果质量：616 g
纵径：12.6 cm
横径：9.2 cm
果形指数：1.4
未成熟果实果皮颜色：暗绿色
成熟果实果皮颜色：亮黄 / 淡黄色
果颈：无
果基：平
果顶：平
果实小果能否剥离：可剥离
果眼外观：突起 / 隆起
果眼深度：深
果眼排列方式：左旋或右旋
果瘤：无
果肉颜色：黄色
果实香味：清香 / 微香
果实风味：甜酸
果肉质地：脆 / 爽脆
果实外观综合评价：中
果实品质综合评价：中
果肉可溶性固形物含量：14.3%
果实成熟特性：中熟
成熟期的一致性：一致
丰产性：低产

Chapter 3 Pineapple Germplasm Resources from Abroad

Queen No. 1 from Vietnam

Code: ACCESSION000043
Name: Queen No. 1 from Vietnam
Place of origin: Vietnam
Resource type: introduced variety
Main ways of consumption: fresh or processing
Initially introduction place: SSCRI, CATAS introduced from Vietnam in 2009
Plant posture: spreading
Planting time: late February in 2021
Open heart stage: mid-February in 2022
Vegetative growth period: approx. 360 d
Initial flowering time: mid-March in 2022
Flowering period: approx. 24 d
Fruit maturation time: early June in 2022
Fruit developing period: approx. 80 d
Fruit shape: long cylindrical
Single fruit weight: 616 g
Longitudinal diameter: 12.6 cm
Vertical diameter: 9.2 cm
Fruit shape index: 1.4
Immature fruit peel color: dark green
Mature fruit peel color: bright yellow/light yellow
Fruit neck: absent
Fruit base: flat
Fruit top: flat
Fruitlets adhesion situation: separable
Fruit eyes appearance: convex/embossed
Fruit eyes depth: deep
Fruit eyes arrangement: levorotation or dextrorotation
Fruit tumors: absent
Flesh color: yellow
Fruit aroma: faint/slight scent
Fruit flavour: sweet-sour
Flesh texture: crisp/crunchy
Fruit appearance comprehensive evaluation: medium
Fruit quality comprehensive evaluation: medium
Soluble solids content in flesh: 14.3 %
Fruit ripening characteristics: medium mature
Maturity consistency: uniform
Productivity: low

植株
Plant

现红
Open heart

花
Flower

花序
Inflorescence

第三章 国外菠萝种质资源
Chapter 3 Pineapple Germplasm Resources from Abroad

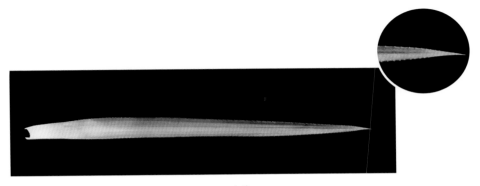

叶片
Leaf

带冠芽果
Fruit with crown bud

冠芽叶刺
Leaf spines in crown bud

果实
Fruit

果实纵切
Fruit longitudinal section

果实横切
Fruit transverse section

越南引无刺卡因 2 号

编号：ACCESSION000042
种质名称：越南引无刺卡因 2 号
原产地：越南
资源类型：引进品种
主要用途：鲜食或加工
种质来源地：2009 年中国热带农业科学院南亚热带作物研究所从越南引进
植株姿态：开张
定植期：2021 年 2 月下旬
现红期：2022 年 2 月下旬
营养生长期：约 365 d
初花期：2022 年 3 月下旬
花开放时间：约 26 d
成熟期：2022 年 6 月中旬
果实发育期：约 110 d
果实形状：长圆筒形
单果质量：1239 g
纵径：15.8 cm
横径：11.1 cm
果形指数：1.4

未成熟果实果皮颜色：暗绿色
成熟果实果皮颜色：暗黄/深黄色
果颈：有
果基：平
果顶：浑圆
果实小果能否剥离：不可剥离
果眼外观：扁平或微凹
果眼深度：较浅
果眼排列方式：左旋或右旋
果瘤：少
果肉颜色：淡黄色
果实香味：清香/微香
果实风味：清甜
果肉质地：脆/爽脆
果实外观综合评价：中
果实品质综合评价：中
果肉可溶性固形物含量：12.7%
果实成熟特性：中晚熟
成熟期的一致性：不一致
丰产性：丰产

Smooth Cayenne No. 2 from Vietnam

Code: ACCESSION000042
Name: Smooth Cayenne No. 2 from Vietnam
Place of origin: Vietnam
Resource type: introduced variety
Main ways of consumption: fresh or processing
Initially introduction place: SSCRI, CATAS introduced from Vietnam in 2009
Plant posture: spreading
Planting time: late February in 2021
Open heart stage: late February in 2022
Vegetative growth period: approx. 365 d
Initial flowering time: late March in 2022
Flowering period: approx. 26 d
Fruit maturation time: mid-June in 2022
Fruit developing period: approx. 110 d
Fruit shape: long cylindrical
Single fruit weight: 1239 g
Longitudinal diameter: 15.8 cm
Vertical diameter: 11.1 cm
Fruit shape index: 1.4
Immature fruit peel color: dark green
Mature fruit peel color: dark yellow/deep yellow
Fruit neck: present
Fruit base: flat
Fruit top: perfectly round
Fruitlets adhesion situation: inseparable
Fruit eyes appearance: flat or slightly concave
Fruit eyes depth: relatively shallow
Fruit eyes arrangement: levorotation or dextrorotation
Fruit tumors: few
Flesh color: pale yellow
Fruit aroma: faint/slight scent
Fruit flavour: mildly sweet
Flesh texture: crisp/crunchy
Fruit appearance comprehensive evaluation: medium
Fruit quality comprehensive evaluation: medium
Soluble solids content in flesh: 12.7 %
Fruit ripening characteristics: medium-late mature
Maturity consistency: nonuniform
Productivity: productive

菠萝种质资源图谱 （上册）
Pineapple Germplasm Resources Map (Volume 1)

植株
Plant

现红
Open heart

花
Flower

花序
Inflorescence

第三章 国外菠萝种质资源
Chapter 3 Pineapple Germplasm Resources from Abroad

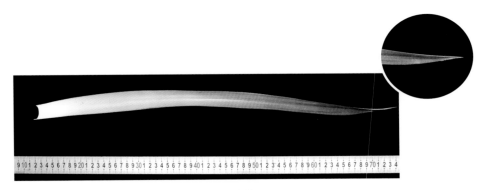

叶片
Leaf

带冠芽果
Fruit with crown bud

冠芽叶刺
Leaf spines in crown bud

果实
Fruit

果实纵切
Fruit longitudinal section

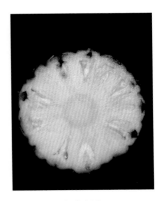

果实横切
Fruit transverse section

肯尼亚引无刺卡因 1 号

编号：ACCESSION000040
种质名称：肯尼亚引无刺卡因 1 号
原产地：肯尼亚
资源类型：引进品种
主要用途：鲜食或加工
种质来源地：2007 年中国热带农业科学院南亚热带作物研究所从肯尼亚引进
植株姿态：开张
定植期：2021 年 2 月下旬
现红期：2022 年 3 月下旬
营养生长期：约 400 d
初花期：2022 年 4 月上旬
花开放时间：约 19 d
成熟期：2022 年 7 月上旬至 8 月下旬
果实发育期：约 115 d
果实形状：圆锥形
单果质量：1257 g
纵径：14.5 cm
横径：11.4 cm
果形指数：1.3

未成熟果实果皮颜色：绿色
成熟果实果皮颜色：金黄色
果颈：无
果基：平
果顶：钝圆
果实小果能否剥离：不可剥离
果眼外观：微隆起
果眼深度：深
果眼排列方式：左旋
果瘤：多
果肉颜色：黄色
果实香味：清香/微香
果实风味：酸甜
果肉质地：粗糙
果实外观综合评价：中
果实品质综合评价：差
果肉可溶性固形物含量：11.5%
果实成熟特性：中晚熟
成熟期的一致性：不一致
丰产性：丰产

Smooth Cayenne No. 1 from Kenya

Code: ACCESSION000040
Name: Smooth Cayenne No. 1 from Kenya
Place of origin: Kenya
Resource type: introduced variety
Main ways of consumption: fresh or processing
Initially introduction place: SSCRI, CATAS introduced from Kenya in 2007
Plant posture: spreading
Planting time: late February in 2021
Open heart stage: late March in 2022
Vegetative growth period: approx. 400 d
Initial flowering time: early April in 2022
Flowering period: approx. 19 d
Fruit maturation time: from early July to late August in 2022
Fruit developing period: approx. 115 d
Fruit shape: conical
Single fruit weight: 1257 g
Longitudinal diameter: 14.5 cm
Vertical diameter: 11.4 cm
Fruit shape index: 1.3
Immature fruit peel color: green
Mature fruit peel color: golden yellow
Fruit neck: absent
Fruit base: flat
Fruit top: bluntly round
Fruitlets adhesion situation: inseparable
Fruit eyes appearance: slightly embossed
Fruit eyes depth: deep
Fruit eyes arrangement: levorotation
Fruit tumors: many
Flesh color: yellow
Fruit aroma: faint/slight scent
Fruit flavour: sour-sweet
Flesh texture: coarse
Fruit appearance comprehensive evaluation: medium
Fruit quality comprehensive evaluation: poor
Soluble solids content in flesh: 11.5 %
Fruit ripening characteristics: medium-late mature
Maturity consistency: nonuniform
Productivity: productive

菠萝种质资源图谱（上册）
Pineapple Germplasm Resources Map (Volume 1)

植株
Plant

现红
Open heart

花
Flower

花序
Inflorescence

第三章 国外菠萝种质资源
Chapter 3 Pineapple Germplasm Resources from Abroad

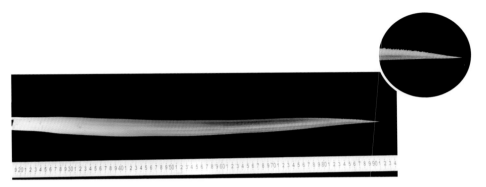

叶片
Leaf

带冠芽果
Fruit with crown bud

冠芽叶刺
Leaf spines in crown bud

果实
Fruit

果实纵切
Fruit longitudinal section

果实横切
Fruit transverse section

法国野生种

编号：ACCESSION000103
种质名称：法国野生种
原产地：法国
资源类型：引进品种
主要用途：育种
种质来源地：2016 年中国热带农业科学院南亚热带作物研究所从法国引进
植株姿态：开张
定植期：2021 年 3 月上旬
现红期：2022 年 2 月下旬
营养生长期：约 355 d
初花期：2022 年 3 月中旬
花开放时间：约 17 d
成熟期：2022 年 6 月中旬
果实发育期：约 90 d
果实形状：长圆筒形
单果质量：35 g
纵径：5.0 cm
横径：3.6 cm
果形指数：1.4

未成熟果实果皮颜色：淡黄 / 黄绿色
成熟果实果皮颜色：黄红色
果颈：无
果基：平
果顶：平
果实小果能否剥离：可剥离
果眼外观：突起 / 隆起
果眼深度：浅
果眼排列方式：左旋
果瘤：无
果肉颜色：白色
果实香味：无
果实风味：微涩
果肉质地：粗糙
果实外观综合评价：中
果实品质综合评价：差
果肉可溶性固形物含量：7.2%
果实成熟特性：中熟
成熟期的一致性：一致
丰产性：低产

Wild Type from France

Code: ACCESSION000103
Name: Wild Type from France
Place of origin: France
Resource type: introduced variety
Main ways of consumption: breeding
Initially introduction place: SSCRI, CATAS introduced from France in 2016
Plant posture: spreading
Planting time: early March in 2021
Open heart stage: late February in 2022
Vegetative growth period: approx. 355 d
Initial flowering time: mid-March in 2022
Flowering period: approx. 17 d
Fruit maturation time: mid-June in 2022
Fruit developing period: approx. 90 d
Fruit shape: long cylindrical
Single fruit weight: 35 g
Longitudinal diameter: 5.0 cm
Vertical diameter: 3.6 cm
Fruit shape index: 1.4
Immature fruit peel color: light yellow/ yellowish green
Mature fruit peel color: yellowish red
Fruit neck: absent
Fruit base: flat
Fruit top: flat
Fruitlets adhesion situation: separable
Fruit eyes appearance: convex/ embossed
Fruit eyes depth: shallow
Fruit eyes arrangement: levorotation
Fruit tumors: absent
Flesh color: white
Fruit aroma: no scent
Fruit flavour: astringent
Flesh texture: coarse
Fruit appearance comprehensive evaluation: medium
Fruit quality comprehensive evaluation: poor
Soluble solids content in flesh: 7.2 %
Fruit ripening characteristics: medium mature
Maturity consistency: uniform
Productivity: low

菠萝种质资源图谱 （上册）
Pineapple Germplasm Resources Map (Volume 1)

植株
Plant

现红
Open heart

花
Flower

花序
Inflorescence

第三章 国外菠萝种质资源
Chapter 3　Pineapple Germplasm Resources from Abroad

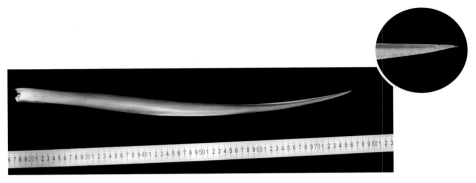

叶片
Leaf

带冠芽果
Fruit with crown bud

冠芽叶刺
Leaf spines in crown bud

果实
Fruit

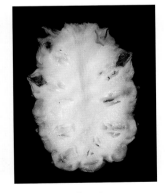

果实纵切
Fruit longitudinal section

果实横切
Fruit transverse section

参考文献　Reference

陈业渊，2005. 热带、南亚热带果树种质资源描述规范 [M]. 北京：中国农业出版社.

李渊林，曾小红，孙光明，2008. 国外菠萝品种资源 [J]. 世界农业，345(1)：55-58.

李渊林，孙光明，2007. 台湾主要商业凤梨品种 [J]. 热带农业科学，27(6)：43-45.

徐迟默，杨连珍，2007. 菠萝科技研究进展 [J]. 华南热带农业大学学报，13(1): 24-29.

中华人民共和国农业部，2015. NY/T 2813—2015　热带作物种质资源描述规范　菠萝 [S]. 北京：中国农业出版社.

D'EECKENBRUGGE G C, LEAL F, DUVAL M F, 1997. Germplasm resources of pineapple[J/OL]. Horticultural Reviews. https://doi.org/10.1002/9780470650660.ch5.